T0200398

Newcomb's Problem

Newcomb's Problem is a controversial paradox of decision theory. It is easily explained and easily understood, and there is a strong chance that most of us have actually faced it in some form or other. And yet it has proven as thorny and intractable a puzzle as much older and better-known philosophical problems of consciousness, skepticism and fatalism. It brings into very sharp and focused disagreement several long-standing philosophical theories concerning practical rationality, the nature of free will and the direction and analysis of causation. This volume introduces readers to the nature of Newcomb's Problem, and ten chapters by leading scholars present the most recent debates around the Problem and analyze its ramifications for decision theory, metaphysics, philosophical psychology and political science. Their chapters highlight the status of Newcomb's Problem as a live and continuing issue in modern philosophy.

Arif Ahmed is Reader in Philosophy at the University of Cambridge. He is the author of *Saul Kripke* (2007), *Wittgenstein's 'Philosophical Investigations': A Reader's Guide* (2010), and *Evidence, Decision and Causality* (2014), and is the editor of *Wittgenstein's 'Philosophical Investigations': A Critical Guide* (2010).

Classic Philosophical Arguments

Over the centuries, a number of individual arguments have formed a crucial part of philosophical enquiry. The volumes in this series examine these arguments, looking at the ramifications and applications which they have come to have, the challenges which they have encountered, and the ways in which they have stood the test of time.

Titles in the Series

The Prisoner's Dilemma
Edited by Martin Peterson
The Original Position
Edited by Timothy Hinton
The Brain in a Vat
Edited by Sanford C. Goldberg
Pascal's Wager
Edited by Paul Bartha and Lawrence Pasternack
Ontological Arguments
Edited by Graham Oppy
Newcomb's Problem
Edited by Arif Ahmed

Newcomb's Problem

Edited by

Arif Ahmed
University of Cambridge

CAMBRIDGE
UNIVERSITY PRESS

CAMBRIDGE
UNIVERSITY PRESS

University Printing House, Cambridge CB2 8BS, United Kingdom

One Liberty Plaza, 20th Floor, New York, NY 10006, USA

477 Williamstown Road, Port Melbourne, VIC 3207, Australia

314–321, 3rd Floor, Plot 3, Splendor Forum, Jasola District Centre, New Delhi – 110025, India

79 Anson Road, #06–04/06, Singapore 079906

Cambridge University Press is part of the University of Cambridge.

It furthers the University's mission by disseminating knowledge in the pursuit of education, learning, and research at the highest international levels of excellence.

www.cambridge.org
Information on this title: www.cambridge.org/9781107180277
DOI: 10.1017/9781316847893

© Cambridge University Press 2018

First published 2018

Printed and bound in Great Britain by Clays Ltd, Elcograf S.p.A.

A catalogue record for this publication is available from the British Library.

Library of Congress Cataloging-in-Publication Data
Names: Ahmed, Arif, editor.
Title: Newcomb's problem / edited by Arif Ahmed (University of Cambridge).
Description: Cambridge ; New York, NY : Cambridge University Press, 2018. |
    Series: Classic philosophical arguments | Includes bibliographical references.
Identifiers: LCCN 2018029219 | ISBN 9781107180277 (hardback) | ISBN 9781316632161 (pbk.)
Subjects: LCSH: Game theory. | Choice (Psychology) | Decision making.
Classification: LCC QA269 .N487 2018 | DDC 519.5/42–dc23
LC record available at https://lccn.loc.gov/2018029219

ISBN 978-1-107-18027-7 Hardback
ISBN 978-1-316-63216-1 Paperback

Contents

Contributors

Arif Ahmed is Reader in Philosophy at the University of Cambridge.

Chrisoula Andreou is Professor of Philosophy at the University of Utah.

José Luis Bermúdez is Professor of Philosophy and Samuel Rhea Gammon Professor in Liberal Arts at Texas A&M University.

Melissa Fusco is Assistant Professor of Philosophy at Columbia University.

Robert Grafstein is the Georgia Athletic Association Professor of Political Science at the University of Georgia.

Preston Greene is Assistant Professor of Philosophy at NTU Singapore.

James M. Joyce is the C. H. Langford Collegiate Professor of Philosophy at the University of Michigan.

Yang Liu is Research Fellow in the Faculty of Philosophy at the University of Cambridge.

Huw Price is the Bertrand Russell Professor of Philosophy at the University of Cambridge.

Robert Stalnaker is Professor of Philosophy at the Massachusetts Institute of Technology.

Reuben Stern is a Postdoctoral Fellow at the Munich Centre for Mathematical Philosophy.

Acknowledgments

I wish to thank all of the contributors to this volume for the care that they took over their contributions and for the patience with which they all dealt with my queries and comments. I am grateful to John Jacobs for his meticulous copy-editing work and to Stephen Duxbury for preparing the index. I am once again grateful to my editor at CUP, Hilary Gaskin, for her own patience and for her continued support. I prepared this volume in part during a period of sabbatical leave, and I am grateful to the Faculty of Philosophy at the University of Cambridge, and to Gonville and Caius College, Cambridge, for granting me this time. Some of the preparation took place whilst I held a visiting position at the Department of Linguistics and Philosophy at the Massachusetts Institute of Technology, and I am grateful to that institution, and especially to Jack Spencer and Christine Graham for their support and hospitality.

Introduction

This introduction sets out what Newcomb's Problem is, why it matters, and some things people have said about it. The appendix sets out some formal details of decision theory insofar as these are relevant to Newcomb's Problem.

1 What It Is

1.1 Nozick's Original Version

Credit for Newcomb's Problem should arguably go to Michael Dummett.[1] But Robert Nozick's eponymous 1969 paper is what set off the enormous debate that followed. Nozick states that he learnt the problem from the physicist William H. Newcomb of the Livermore Laboratory. Nozick puts it as follows.

> **Standard Newcomb**
> You must choose between taking (and keeping the contents of) (i) an opaque box now facing you or (ii) that same opaque box *and* a transparent box next to it containing $1000. Yesterday, a being with an excellent track record of predicting human behaviour in this situation made a prediction about your choice. If it predicted that you would take only the opaque box ('one-boxing'), it placed $1M in the opaque box. If it predicted that you would take both ('two-boxing'), it put nothing in the opaque box.[2]

He goes on: 'To almost everyone it is perfectly clear and obvious what should be done. The difficulty is that these people seem to divide almost evenly on the problem, with large numbers thinking that the opposing half is just being silly'.

---

[1] This is Dummett's problem of the dancing chief (Dummett 1964). One reason to hesitate over this attribution is that in Newcomb's Problem it is stipulated that acts are causally irrelevant to correlated states, whereas the central question of Dummett's paper is over whether this is even possible.

[2] Nozick 1969: 207.

That disagreement matters in part because *we* may face versions of the problem. I'll discuss these at section 2.1. But it also matters because there are arguments for either side, resting on principles that until 1969 had seemed harmonious as well as compelling. Put very simply, these are as follows:

Causal Principle: A rational agent does what she thinks will *cause* her to realize her aims.

Evidential Principle: A rational agent does what constitutes her best *evidence* that she will realize her aims.

The Causal Principle seems to recommend two-boxing. You can't now *make* any difference to the contents of the opaque box, which were settled yesterday. Two-boxing therefore guarantees an extra $1K. The Evidential Principle seems to recommend one-boxing. One-boxing is, and two-boxing is not, excellent evidence that you are about to get $1M. So the Causal Principle and the Evidential Principle cannot *both* be right. I'll discuss these principles more formally at section 2.2.

1.2 General Form of the Problem

The problem invokes three features that are common to all decision problems – acts, outcomes and states; and four that are specific to it – stochastic dependence, causal independence and two kinds of dominance. I'll describe these in turn.

If you are choosing what to do, then your choice is between *acts*. Their *outcomes* are the possible consequences that matter to you. What outcome an act obtains depends on the *state* of nature at the time. In *Standard Newcomb* you choose between the acts of one-boxing and two-boxing. The outcome is monetary and depends on what you were predicted to choose, the latter being the state. Table 1 summarizes all this.

The column headings represent the possible states S_1 and S_2, the row headings represent your options A_1 and A_2, and the interior of the table

Table 1: Standard Newcomb

	S_1: *Predicted A_1*	S_2: *Predicted A_2*
A_1: **Take only the opaque box**	$1M	0
A_2: **Take both boxes**	$1M + $1K	$1K

indicates the outcome, in terms of your payoffs, in each of the four act/state combinations. These are the general features of the problem.

Let me turn to its specific features. First, there is *stochastic dependence* between act and state. When anyone takes only the opaque box, the predictor has almost always predicted this; when anyone takes both boxes, the predictor has almost always predicted *this*. This has two consequences. (a) One-boxers almost always end up millionaires, and two-boxers almost never do. (b) You are very confident that *you* will end up a millionaire if and only if you now take only the opaque box.[3]

Second, states are *causally independent* of acts. Whether you take one box or two *makes* no difference to the prediction. This combination of causal independence with stochastic dependence illustrates the saying that correlation is not causation. There is, e.g., a correlation between weather forecasts and subsequent weather, but weather forecasts have no causal influence on subsequent weather events, nor could any weather event have any (retroactive) influence on prior forecasts of it. The correlation exists because forecasts of weather and actual weather are effects of a common cause, i.e., the atmospheric conditions, etc., that precede both. In Newcomb's Problem, choices might similarly be correlated with predictions because they share a common cause, for instance, some previous state of the agent's brain.

The third feature is that one option *dominates* the other. Given either state – the state of predicted one-boxing, or of predicted two-boxing – you are $1K better off if you two-box. I'll call this feature *horizontal dominance*.

The final feature is that the worst outcome in one state (that of predicted one-boxing) is better than the best outcome in the other. If the true state is S_1, then whatever you do, you are guaranteed a better outcome than you could possibly get in S_2. Right or wrong, a prediction of one-boxing makes you much better off than a prediction of two-boxing could. I'll call this feature *vertical super-dominance*.

These four specific features of Newcomb's Problem are responsible for the tension that it creates. Stochastic dependence and vertical super-dominance seem to rationalize one-boxing. After all, almost everyone who one-boxes ends up a millionaire, and almost nobody who two-boxes does, including people who have reasoned just as you might be reasoning now. Why expect to

---

[3] Strictly speaking, neither point logically *follows* from the assumption that the predictor has a good track record; the reasoning involves inductive inference from the latter. In Newcomb's Problem as in everyday life, we waive skeptical concerns about induction. We take it for granted that the predictor's past track record *is* strong evidence (a) of his future track record and (b) that he predicted you right on this occasion.

buck this trend? Causal independence and horizontal dominance seem to rationalize two-boxing. After all, either the $1M is already in the opaque box, or the opportunity to put it there is past, and nothing that you do can make a difference to whether it is there; and either way you are better off two-boxing.

2 Why It Matters

Newcomb's Problem is an interesting intellectual exercise, but so are many other things that have attracted less expenditure of thought and time. I think there are two reasons for the intense interest that *this* problem continues to provoke. (i) Its abstract structure seems to apply to cases that really do, or easily could, arise in real life. (ii) It motivated a profound shift in the way we think about rational choice. I'll take these points in turn.

2.1 Realistic Newcomb Problems

Here are two Newcomb Problems that could easily be, or probably are, real.

(a) *Fisher smoking case.*[4] Suppose that what explains the correlation between smoking and lung disease is not (as everyone now thinks) that smoking causes lung disease, but rather that both have a common cause: an innate predisposition towards lung diseases that also, and separately, predisposes its bearers to smoke. Suppose you are wondering whether to smoke, but you don't know whether you have the predisposition. You know that you would like smoking, but you like good health very much more.

   The decision involves all four factors that distinguish a Newcomb Problem. Smoking and lung disease are stochastically related but causally independent. Smoking dominates non-smoking: whether or not you have the predisposition, you are better off smoking than not. And because you care a lot more about lung disease than about smoking, the absence of the predisposition super-dominates its presence. In this version, not smoking corresponds to one-boxing and smoking corresponds to two-boxing.

(b) *Voting in large elections.*[5] In a large election it is almost certainly true of you, as of any individual voter, that your vote won't affect the outcome. On the other hand, you might think your voting is symptomatic of whether others like you, in particular supporters of your candidate, will

---

[4] Jeffrey 1983: 15.   [5] Quattrone and Tversky 1986: 48–57.

vote. If you expect turn-out to be decisive, your voting for your preferred candidate may be evidence of your preferred outcome.

If so, the case is a good approximation to Newcomb's Problem: good enough, that is, to raise the same problems. Let S_1 and S_2 be possible outcomes of the election – either your candidate wins or she does not – and let A_1 and A_2 be the options of voting for your candidate and not voting at all. The problem satisfies causal independence, nearly enough: it is practically certain that your vote *makes* no difference to the outcome of the election.[6] It satisfies horizontal dominance: given that your candidate wins, or given that she doesn't, you are better off not incurring the small opportunity cost of voting. And it satisfies vertical super-dominance, if it matters greatly to you that your candidate wins.

It is less obvious that elections involve stochastic dependence, but there is evidence that they do. From a purely statistical perspective, this is not surprising: if we consider all choices whether to vote, amongst Republican supporters from every US Presidential election that took place in the twentieth century, we should expect a correlation between a choice's having been to vote (rather than abstain) and the Republican candidate's having won. More importantly, this correlation has a subjective counterpart. Many people *do* think of their own choices as symptomatic of the choices of people like them, including in the context of large elections.[7] Any such person therefore faces a real-life Newcomb Problem in any large election in which he (i) can vote and (ii) has a strong interest. In this version of the problem, voting corresponds to one-boxing and not voting corresponds to two-boxing. (For more discussion, see the chapter by Grafstein in this volume.)

Those are two examples of Newcomb's Problem. The literature notes many others.

(c) The choice between vice and virtue in the context of Calvinist pre-destination.[8]

(d) Macroeconomic policy choice in the context of rational expectations[9] (but see the chapter by Bermúdez in this volume).

---

[6] For instance: in the UK since 1832, there have been five elections for Parliamentary representatives, out of approximately 30,000 such, in which the margin of victory has been zero or in single figures. This gives a frequency of about 0.05% of cases in which the margin of victory was in single figures.

[7] For evidence that they (i) think this and (ii) do so reasonably, see Ahmed 2014a section 4.6.

[8] Resnik 1987: 111; Ahmed 2014a: 9ff.; for historical details, see Weber 1992; Tawney 1998.

[9] Frydman, O' Driscoll and Schotter 1982.

(e) The choice whether to engage in some mildly unpleasant activity that is symptomatic of cardiac health.[10]

(f) The choice whether to smoke, when present smoking indicates future smoking.[11]

(g) Bets about experiments involving non-causal quantum correlations.[12]

(h) Choices in the Libet experiment, where experimenters can predict the agent's decision before she becomes consciously aware of it.[13]

(i) Bets about the prior state of the world in the context of determinism.[14]

(j) Decisions made by autonomous vehicles in an environment containing many similar agents.[15]

(k) Prisoners' Dilemma (from game theory) also realizes Newcomb's Problem, *if* each prisoner is confident enough that both reason alike[16] (but see the chapter by Bermúdez in this volume).

So despite its typically science-fictional presentation, the basic structure of Newcomb's Problem is arguably realistic, and its realizations may be very widespread.

2.2 Causal and Evidential Decision Theory

Probably the most important philosophical insight to have arisen from Newcomb's Problem is the distinction between two systematic ways of thinking about practical rationality. Perhaps the best way to understand the difference between these is in terms of two possible responses to the ancient philosophical problem of *fatalism*.

That problem itself arises from an overextension of a natural principle of rationality, which we may call the principle of *dominance*, a very simple version of which we may write as follows:

Dominance: For any two acts A_1 and A_2, if for each state the outcome of A_2 is better for you in every state than the outcome of A_1, then it is rational to choose A_2 over A_1.

Dominance looks perfectly reasonable: if, for instance, investing in gold gets you a better return than investing in land if the Republicans control the Senate after the next election, and gold also gets a better return than land if they do not, then it is sensible to invest in gold rather than land.

---

[10] Quattrone and Tversky 1986: 41–8.   [11] Monterosso and Ainslie 1999; Ahmed forthcoming.
[12] Cavalcanti 2010; Ahmed and Caulton 2014.   [13] Slezak 2013.   [14] Ahmed 2014a section 5.2.
[15] Meyer, Feldmaier and Shen 2016.   [16] Lewis 1979.

But the fatalist argument shows that in the absence of restrictions, the principle of Dominance leads to absurd consequences. Here is Cicero's report of one such case:

> So their [the Stoics'] argument goes: 'If you are destined to recover from this illness, whether you were to call in a doctor or not, you would recover; furthermore, if you are destined not to recover from this illness, whether you were to call in a doctor or not, you would not recover—and either one or the other is destined to happen; therefore it doesn't matter if you call in a doctor.'[17]

Dominance seems to validate this argument, and if calling a doctor carries any cost then it appears to recommend *not* calling a doctor. This is apparent if we lay out the acts, the states, and notional values for the outcomes as follows:

Table 2: Fatalism

	S_1: You recover	S_2: You don't
A_1: Call the doctor	5	0
A_2: Don't	6	1

In either state you are better off having not called the doctor than having called the doctor. Dominance seems to recommend not calling the doctor in this situation, however ill you are. More generally it seems to recommend the fatalist strategy of never taking a costly means to *any* end, however desirable.

Intuitively, the flaw in this reasoning is that it overlooks any *connection* between act and state. More specifically, we might expect dominance to fail when one of the acts in some sense makes the state more likely than does the other.

But this diagnosis, whilst correct, leaves room for two interpretations of 'making more likely'. We might say (i) that Dominance fails only if (as it seems to the agent) the acts have a *causal* influence on the state; or we might say (ii) that Dominance fails if the acts are (again from the agent's perspective) *evidentially* relevant to the state. Both (i) and (ii) are enough to undercut the fatalist argument, since calling the doctor is *both* a cause of, *and* evidence for, your recovery. But generalizing these diagnoses into a principle of action

---

[17] *De Fato* 28–9.

gives rise to theories of rationality that differ elsewhere, and in particular over Newcomb's Problem.

The natural generalization of (i) gives rise to *Causal Decision Theory* (CDT). This theory of rationality has various formal realizations that are not precisely equivalent; but what they all have in common is the idea that the rational act is whichever available one is most likely to *cause* what you want to happen.[18] The natural generalization of (ii) gives rise to *Evidential Decision Theory* (EDT), according to which the rational act is whichever available one is the best evidence of what you want to happen.[19]

EDT and CDT agree that you should call the doctor in *Fatalism*. And they agree wherever one's options are between acts that are evidence for exactly those states that they causally promote. But they do not agree over cases where one's acts are evidence for states that they do not causally promote, and this is exactly the situation in Newcomb's Problem. One-boxing is evidence that you will get $1M because it is evidence of the state in which you were predicted to one-box; EDT therefore recommends one-boxing. Two-boxing brings it about that you are $1K richer than you would otherwise have been; CDT therefore recommends two-boxing. Newcomb's Problem measures the distance between the thought that rational choice must pay special attention to *causal* dependences of states on acts and the thought that it need only be sensitive to the extent to which acts are *evidence* of states. The dispute over Newcomb's Problem is therefore one aspect of the more general epistemological question concerning the place of causality itself in our conception of the universe. (For technical details of EDT and CDT, see the Appendix to this Introduction. For discussion of alternative versions of CDT, see the chapter by Stern. For the connections between CDT and game theory, see the chapter by Stalnaker.)

3 The Debate Since 1969

In light of its apparently wide application, one might have expected discussion of Newcomb's Problem to have flourished in all those branches of science that

---

[18] The theory originated with Stalnaker 1972. For other versions, see Lewis 1981; Sobel 1986; Joyce 1999. All of these theories agree (a) that causal beliefs play a central role in rational decision-making; (b) that you should two-box in Newcomb's Problem. Spohn's (2012) and Price's (2012) versions of CDT both accept (a), but they reject (b), for different reasons. The Appendix to this Introduction spells out the relatively simple and early version attributed to Gibbard and Harper (1978).

[19] Jeffrey 1965 remains the classic exposition of Evidential Decision Theory. (Jeffrey 1983, the second edition of that book, modifies the theory so that it recommends taking both boxes in Newcomb's Problem – see Jeffrey 1983: 15–25.)

deal with choice, including economics, psychology and political science as well as philosophy. As it happened, professional discussion of the problem in the 1970s and early 1980s was largely conducted amongst philosophers. But there has since the 1980s been increasing (though still hardly mainstream) interest in the problem within these other areas and more lately in robotics and computer science. The following is a necessarily partial summary of some highlights of the philosophical literature.

3.1 Are Newcomb Problems Possible?

The "tickle defense" purports to show that, appearances to the contrary, nobody ever faces real Newcomb Problems. Informally, the idea is as follows. Newcomb's Problem arises only when you think that your act is evidence of a causally independent state. But this can only happen if either the state either itself causes you to act in some way or is a side effect of some prior cause of your act. Either way, this prior cause of your act must be mediated by your *motivations* – your desires and beliefs. At the time of acting, you know what your motivations are. But if you *know* that, you won't regard the act they produce as *further* evidence of its distal cause, nor therefore of the state. All the evidential bearing that your act could have on the state is already available from your known motivations.[20]

Thus, consider the Fisher smoking example described at 2.1(a). Smoking is supposed to indicate a predisposition that causes it. And it's plausible that learning that someone *else* smokes is evidence for you that she has that predisposition. But this is not so clear when it comes to *your own* smoking. If you are predisposed to smoke, then presumably you already like the idea of smoking (you have a "tickle" or urge to smoke), and whether you do is something that you already know. But the predisposition only makes you smoke by making you like the idea, and since you already know about that, your actual choice reveals no more about the presence or absence of the predisposition. From the perspective of the agent herself, smoking is therefore not any sort of evidence of a state that it doesn't cause. The Fisher smoking case is therefore not a Newcomb Problem.

Several issues arise here. We might question the quasi-Cartesian assumption that you know your own motivational state. The contrary idea, that subconscious desires and beliefs can play the same role in motivation as familiar conscious ones, is familiar from Freud, and whatever you think of

---

[20] Eells 1982: ch. 6, 7.

that, you might also think that a degree of muddle about what you think or want is a human imperfection that we cannot simply assume away.[21]

A second assumption of the tickle defense is that your acts only correlate with non-effects of them that are either their own causes or share some common cause with them. But cases of quantum entanglement cast doubt on this assumption, and it is possible to construct quantum cases that create a Newcomb-like clash between EDT and CDT but are immune to the tickle defense.[22]

But there is a second reason to doubt the possibility of Newcomb Problems. It is crucial that the state is *causally* independent of what you now do. But what *is* this causal relation? After all, nobody ever observes a relation of causality between distinct events, in the way that one observes, say, the relation of harmony or discord between distinct musical tones. So what is it? One possible answer, suggested by Berkeley but developed more thoroughly by Menzies and Price, is that A causes B when there is a correlation between an agent's *directly bringing about* A and the occurrence of B.[23]

This makes Newcomb's Problem impossible. We are told that the contents of the opaque box are correlated with what you bring about, i.e., whether you choose to take one box or two. But on the present view, this means that your choice *causes* the opaque box to contain \$1M, or to contain nothing. That contradicts the stipulation that the contents of the opaque box are causally *independent* of what you do.

In response, one might think that there is a strong independent reason to doubt this "agency theory of causation." It may be that the account cannot be generalized to cover all impersonal causal relations without draining it of content.[24] It may be that the idea of an agent is itself causal in some way that makes the theory objectionably circular.[25] And it is a disturbing consequence of the theory that it appears to sacrifice the asymmetry as well as the temporal directedness of causation: since it is doubtless true, whether or not Newcomb's Problems are possible, that human actions have causes with which they are correlated, the theory is committed to saying quite implausibly that human actions are the causes as well as the effects of their own causes. (For further discussion, see the chapter by Price and Liu.)

---

[21] Cf. Lewis 1981a: 311–2.   [22] For details, see Ahmed and Caulton 2014.
[23] Berkeley 1980 [1710]: sect. 25ff.; Menzies and Price 1993.
[24] For this and other criticisms of the agency theory, see Woodward 2003: 123–7.
[25] For discussion of various such objections, see Ahmed 2007.

3.2 Why Ain'cha Rich?

Supposing Newcomb's Problem is possible, what can be said for or against either side? One prominent argument for one-boxing arises from the empirically minded view that the better option is that with the better average return.

Suppose that the predictor anticipates either strategy with 90% accuracy. Then across cases in which the subject takes one box, she gets $1M 90% of the time and nothing 10% of the time, for an average return of $0.9M. And across cases in which she takes both boxes, she gets $1M+$1K 10% of the time and $1K 90% of the time, for an average return of $0.101M. The actual average return to one-boxing is about nine times greater than the average return to two-boxing. This fact, on which all sides must agree, is the premise from which a standard argument, known as 'Why Ain'cha Rich?', infers that one-boxing is rational.

Two-boxers will reply that this comparison is unfair. When you took both boxes, it is not as though you *could* have gotten any more than the $1K that you actually got. The predictor had already foregone the opportunity to put $1M into the opaque box, so $1K was the most that you could have got. The reason that those who take both boxes are not rich is that *those* people *could* not have gotten rich. But still, they did the best that they could. By similar reasoning, those who declined the second box got rich because they, as it happened, could not have *failed* to get rich. But they *could* have gotten *richer*.

Some philosophers allege further that 'Why ain'cha rich?' proves too much. Consider a version of Newcomb's Problem where *both* boxes are transparent. In this case, it seems obviously rational to take both boxes. But people who take both boxes were typically predicted to choose that option and so typically see nothing in the formerly opaque box and get nothing from it. The people who *do* typically end up rich are the very few who decline to take both boxes even in this case. That seems insane, but these "lunatics" could retort, to the rest of us who have faced this problem: "If you're so smart, why aren't *you* rich?" (For further discussion, see the chapters by Ahmed and Greene.)

3.3 Spin-offs and Deliberational Dynamics

Standard Newcomb combines (i) states of which your acts are symptoms but not causes, with (ii) payoffs on which the act that is symptomatic of the inferior state dominates the act that is symptomatic of the superior state. But philosophers have invented a variety of problems that share (i) but not (ii) with *Standard Newcomb*.

Perhaps the most famous of these is *Appointment in Samarra* (or "Death in Damascus"), deriving from a short story of that title by W. Somerset Maugham. In the story, a servant is jostled in a crowded Baghdad marketplace by (he thinks) a woman who makes a threatening gesture. Looking more closely, he sees that it is Death. He immediately borrows his master's horse and rides full tilt to Samarra, where (he thinks) Death will not find him. Death takes up the story: "Then the merchant went down to the marketplace and he saw me standing in the crowd and he came to me and said, 'Why did you make a threating gesture to my servant when you saw him this morning?' 'That was not a threatening gesture', I said, 'it was only a start of surprise. I was astonished to see him in Baghdad, for I had an appointment with him tonight in Samarra.'"

Death works from a highly accurate appointment book assigning a time and a place for each person; a person dies at a time if and only if the book correctly states his location then. Your appointment is tonight, and it is in Baghdad or Samarra. You must choose now between staying in Baghdad and riding to Samarra, to arrive tonight.[26] The following table sets out the options, states and notional payoffs of this problem.

Table 3: Appointment in Samarra

	S_1: Predicts Baghdad	S_2: Predicts Samarra
A_1: **Stay in Baghdad**	0	\$1M
A_2: **Ride to Samarra**	\$1M	0

There is a correlation between your ultimate destination and Death's present location. But the former is supposed to have no *effect* on the latter.

Evidential Decision Theory is indifferent between A_1 and A_2 since both options are equally good, or rather equally bad news: riding to Samarra is a near-certain sign that Death will be waiting for you there; staying put is an equally good sign that he was here all along. Whatever your initial confidence about what is in the appointment book, EDT considers flight no better (or worse) than staying put. Given the story, this fatalism is intuitively plausible, and the case raises no special difficulty for EDT. By contrast, what CDT

---

[26] Gibbard and Harper 1978: 373 with trivial alterations. For something more realistic: imagine that smoking is harmful to you if and only if you possess a gene that predisposes you to smoke.

recommends *does* seem to depend on your initial view of things. If you are more confident that Death has predicted Baghdad, it recommends flight, because that *makes* your survival more likely; if you are more confident that Death has predicted Samarra, then it recommends staying put for the same reason.

It may seem troubling that CDT positively recommends a course of action that you know in advance will be as self-defeating as the alternative. Worse, it seems to generate instability. Suppose you start out confident that Death is in Samarra and so decide to stay put. But having decided that, you grow confident that Death is in Baghdad. You therefore change your mind. But doing so makes you more confident that Death awaits in Samarra. Whatever you decide to do, CDT will advise the alternative. The problem is not that CDT recommends perpetual dithering (which isn't an option) but that CDT seems to undermine its own advice to anyone who takes it.

One promising and fruitful line of response involves *deliberational dynamics*.[27] We model deliberation as follows. Suppose that you follow CDT. But if CDT advises that you choose an act, the effect of this falls short of your carrying out the act; rather it increases your confidence that you will carry it out. The effect of *that* will be to revise your confidence in each state. You then reconsider what to do in accordance with these revised levels of confidence, again using CDT.

This process makes the probability of flight evolve across successive stages of the deliberating self. For instance, suppose that you start out highly confident that Death is in Baghdad. So if you follow CDT, then you become more confident that you'll flee, because now you are more confident that the effect of flight is survival. But since you know Death's appointment book is highly accurate, this now raises your confidence that Death is in Samarra; taking this into account, you now deliberate again.

Deliberational Causal Decision Theory (DCDT) advises the agent to follow CDT at the point where he has taken account of *all* relevant information, including any that arises in this way from the deliberative process itself: that is, at any point at which further deliberation will do nothing to change the agent's mind. Under certain quantitative assumptions, the existence of such a 'fixed point' is guaranteed by the same sort of reason that guarantees a mixed Nash equilibrium in game theory; more generally, a fixed point is any distribution of confidence over the states relative to which CDT (a) is indifferent between all options to which that distribution gives positive probability,

---

[27] Arntzenius 2008; Joyce 2012.

and (b) prefers any such option to any other. At this point of indifference and whatever the agent's initial beliefs, CDT does *not* positively recommend a course of action over alternatives that are no more self-defeating; neither is there any instability. DCDT is, like EDT, indifferent between staying in Baghdad and riding to Samarra. And in Newcomb's Problem, equilibrium exists only at the point where the agent is certain that he will take both boxes; and at this point as at any other, CDT recommends taking both. Still, DCDT does face the objection that in *Psychopath Button* it *permits* pressing, and this can look counterintuitive. (For discussion of DCDT, and of some of these examples, see the chapter by Joyce in this collection. For connections with other decision-theoretic puzzles, see the chapters by Andreou and Fusco.)

4 Conclusion

I have only been able to touch upon a few of the issues that have arisen in nearly 50 years of debate on Newcomb's Problem. But despite all this activity, neither the simple question of what to do in it, nor the larger question of whether EDT or CDT is the best account of practical rationality, show signs of resolution. This may be because the whole antinomy rests on some deep misunderstanding that we are yet to uncover. Or it may be because Newcomb's Problem represents a crux, not only between one-boxing and two-boxing, and not only between EDT and CDT, but between two fundamentally opposed conceptions of the world, of which the dispute over causality is merely a symptom. In any case, the path to truth runs through clear-eyed antagonism, not woolly consensus. The following chapters disagree sharply over all the points discussed above, and many more besides.

Appendix: Evidential Decision Theory and Causal Decision Theory

This section sets out the formal details of decision theory insofar as these are relevant to Newcomb's Problem. Although they are not necessary for an informal understanding of the problem as already set out, they are presupposed in some of the following chapters, and the reader may find it useful to have all of this material collected in one place.

We suppose, as in section 1.2, that there is a set of options, a set of states of the world, and a set of prizes or outcomes. Given an act A and a state S, the outcome that this act/state pair determines is AS. Its value to you is a number $V(AS)$, called its *utility*: for present purposes we may – though in general one should not – treat the utility of an outcome as its dollar value.

You are uncertain about which state obtains. We measure this uncertainty by means of a kind of probability, called a *credence*, which assigns a number $Cr(S)$ between 0 and 1 to each state of the world S. The credences in these states add up to 1. Your credence in a state represents your degree of confidence that it obtains: roughly, we can think of it as reflecting the worst odds at which you would be willing to bet that that state obtains.

Let us index the available acts A_1, A_2,...A_m and the possible states S_1, S_2,...S_n. The basic theory then assigns an *expected utility EU* to each act as follows:

$$EU(A_i) = \sum_{1 \leq j \leq n} Cr(S_j) V(A_i S_j) \tag{1}$$

This quantity is supposed to be a weighted average of the values of the states given the act, where the weights are the probabilities of the states. And the basic theory says that a rational choice from a set of options is any that *maximizes* expected utility.

For instance, suppose that I offer you a bet on this fair coin: you win \$1 if it lands heads and lose 50 cents if it lands tails. Your options are (A_1) betting, or (A_2) not betting. The states are (S_1) that the coin lands heads on its next toss and (S_2) that it lands tails; and $Cr(S_1) = Cr(S_2) = 0.5$. The prizes, measured in dollars, are $V(A_1 S_1) = 1$, $V(A_1 S_2) = -0.5$, $V(A_2 S_1) = 0$ and $V(A_2 S_2) = 0$. Putting these figures into (1) gives an expected utility for betting of 0.25 and an expected utility for not betting of zero. The theory therefore advises you to bet.

As it will be understood here, the theory of rational choice is *normative*, not descriptive. It doesn't attempt to say what people *do* in the face of subjective uncertainty, but rather what it is *rational* to do. As a *descriptive* theory, expected-utility maximization does not fare well, given the famous paradoxes of Allais and Ellsberg;[28] but it seems to have more going for it as a normative theory of rationality.[29]

Problems arise, however, because EU-maximization implies an unrestricted version of the Dominance principle. As we've seen, this gives rise to problems when acts and states interact, as, for instance, in the *Fatalism* example of section 2.2 (see Table 2). If calling a doctor carries any cost, then EU-maximization appears to make the absurd recommendation that you *not* call a doctor. In particular, applying (1) to Table 2 implies that the expected utility of not calling the doctor exceeds that of calling the doctor by 1 unit. The basic theory seems to recommend not calling the doctor in this situation, however ill you are. More generally it seems to recommend the *fatalist* strategy of never taking a costly means to *any* end, however desirable.

Intuitively, the flaw in this reasoning is that it overlooks any *connection* between act and state. More specifically, we might expect (1) to fail when one of the acts in some sense makes the state more likely than does the other. But as we'll now see, there are two interpretations of "making more likely."

Evidential Decision Theory (EDT) amends (1) to reflect, not the probability of the state itself, but rather its probability *given* that you perform the evaluated act. It postulates a generalized notion of value ("news value") that applies indifferently to acts, states and outcomes. The news value of an act A_i is given by:

$$V(A_i) = \sum_j Cr(S_j|A_i) V(A_i S_j) \tag{2}$$

The crucial difference between (1) and (2) is that where (1) has $Cr(S_j)$, (2) has $Cr(S_j|A_i)$. This quantity, called the probability of S_j given A_i, is defined as $\frac{Cr(S_j A_i)}{Cr(A_i)}$, and measures A_i's strength as *evidence* for S_j, for you. That is because $Cr(S|A)$ measures the degree of confidence that you *would* have in S if you were to learn that A is true. For instance, if S is the proposition that your

[28] Allais 1953; Ellsberg 1961.
[29] But see Grafstein's chapter in this volume for more on the normative / descriptive distinction.

lottery ticket is a winner, A_1 is the proposition that you draw from a lottery in which 1 in 10 tickets is a winner and A_2 is the proposition that you draw from a lottery in which 1 in 100 is, then $Cr(S|A_1) = 0.1$ and $Cr(S|A_2) = 0.01$, implying that you regard A_1 as better evidence than A_2 for S. EDT says that a rational choice from a set of options is any that maximizes V: roughly, it is the best *news* that you could give yourself.

It is easy to see how EDT deals with *Fatalism*. Intuitively, your calling a doctor is better news than your not doing this. If, given your symptoms and general state of health, we learn that you called the doctor, we are more optimistic about your recovery than if we had learnt that you didn't. More formally, your confidence of recovery, given that you call a doctor, must greatly exceed your confidence of recovery given that you don't. So $Cr(S_1|A_1) \gg Cr(S_1|A_2)$ and $Cr(S_2|A_1) \gg Cr(S_2|A_2)$. So, given (2) and the values specified in Table 2, the V-score or news value of calling a doctor exceeds that of not calling one.

EDT treats Newcomb's Problem and Fatalism in the same way. Your taking only the opaque box is evidence that it contains \$1M. Your taking both is evidence that it does not. So you have $Cr(S_1|A_1) \gg Cr(S_1|A_2)$ and $Cr(S_2|A_1) \gg Cr(S_2|A_2)$. EDT therefore recommends one-boxing.

Causal Decision Theory (CDT) recommends doing what will best *bring about* what you want, where "bring about" involves *causality* (whatever that is). More formally, given acts A_i, states S_j and outcomes A_iS_j, the *utility* U of an act is:

$$U(A_i) = \sum_j Cr(A_i \Rightarrow S_j) V(A_iS_j) \tag{3}$$

In (3), $\Rightarrow$ is the (causal) subjunctive conditional: $Cr(A_i \Rightarrow S_j)$ represents your confidence that if A_i *were* true then S_j *would* be true. Any difference between $Cr(A_i \Rightarrow S_j)$ and $Cr(A_{i'} \Rightarrow S_j)$ reflects a difference, in your opinion, between the tendencies of A_i and of $A_{i'}$ to cause the truth of S_j. And you have $Cr(A_i \Rightarrow S_j) = Cr(A_{i'} \Rightarrow S_j)$ if and only if you take the obtaining of S_j to be *causally independent* of your choice between A_i and $A_{i'}$. CDT recommends whichever act maximizes U from amongst those available.

CDT therefore *agrees* with EDT that a rational person facing *Fatalism* will call a doctor; but it *disagrees* with EDT over Newcomb's Problem. Let me take these points in turn. In *Fatalism* the crucial fact, as CDT sees it, is that calling a doctor can help *bring about* recovery. Formally, this is reflected in the facts that $Cr(A_1 \Rightarrow S_1) > Cr(A_2 \Rightarrow S_1)$ and $Cr(A_2 \Rightarrow S_2) > Cr(A_1 \Rightarrow S_2)$. If this difference in causal efficacy is in your view large enough, you will if rational call the doctor, according to (3) and U-maximization.

But in Newcomb's Problem, your choice has no effect on any prediction of it. All you can *bring about* in Newcomb's Problem is that you get the extra $1K in the transparent box; since this $1K is something that you want, doing so is obviously rational. Formally: in Table 1, the state is causally independent of your act. So $Cr(A_1 \Rightarrow S_1) = Cr(A_2 \Rightarrow S_1)$ and $Cr(A_1 \Rightarrow S_2) = Cr(A_2 \Rightarrow S_2)$. Plugging these assumptions into (3) implies $U(A_2) = U(A_1) + K$, so CDT recommends taking both boxes.[30]

---

[30] Causal Decision Theory as explained here is a simplified version of that presented in Gibbard and Harper 1978; the theory itself originated with Stalnaker 1972. For other versions see Lewis 1981a; Sobel 1986; Joyce 1999.

1   Does Newcomb's Problem Actually Exist?

José Luis Bermúdez

As Newcomb's Problem (henceforth: NP) is standardly presented, you are faced with a Predictor in whose predictive reliability you have great confidence. There are two boxes in front of you – one opaque and one transparent. You have to choose between taking just the opaque box (*one-boxing*) and taking both boxes (*two-boxing*). You can see that the transparent box contains $1,000. The Predictor informs you that the opaque box may contain $1,000,000 or it may be empty, depending on how she has predicted you will choose. The opaque box contains $1,000,000 if the Predictor has predicted that you will take only the opaque box. But if the Predictor has predicted that you will take both boxes, the opaque box is empty. This setup yields the payoff table depicted in Table 1.

Table 1

	The Predictor has predicted two-boxing and so the opaque box contains $1,000,000	The Predictor has predicted one-boxing and so the opaque box is empty
Take just the opaque box	$1,000,000	$0
Take both boxes	$1,001,000	$1,000

The principal reason why NP has been so widely discussed is that it appears to reveal a conflict between two independently plausible but (in this case) incompatible decision-making principles.

Dominance. There are two circumstances. Either there is $1,000,000 in the opaque box or there is not. In either case, the payoff is higher for a two-boxer than a one-boxer. So, dominance reasoning prescribes two-boxing rather than one-boxing.

Maximizing expected utility. Suppose you assign a degree of belief of 0.99 to the Predictor's reliability (and suppose that expected utility in this case coincides with expected monetary value). Then the expected utility of one-boxing is 0.99 x \$1,000,000 = \$990,000, while the expected utility of two-boxing is (0.99 x \$1,000) + (0.01 x \$1,001,000) = \$11,000. So, a rational agent maximizing expected utility will take just the opaque box.[1]

Plainly, both principles cannot be applied in NP. The solution proposed by many philosophers and some decision theorists is to rethink the principle of maximizing expected utility.[2] What NP reveals, on this view, is the inadequacy of a purely evidential approach to decision theory. As decision theory is standardly formulated, expected utility is calculated relative to the probabilities of the different outcomes conditional upon one of the available actions being performed. Instead, it is suggested, calculations of expected utility should be based on probability calculations that track causal relations between actions and outcomes, as opposed simply to probabilistic dependence relations between actions and outcomes. Or, in other words, Evidential Decision Theory (EDT) should be replaced by Causal Decision Theory (CDT).

There is a natural worry with this. Thought experiments have a long-standing and important role to play not just in philosophy, but also in science. Think of Newton's bucket and Schrödinger's cat, for example. But still, one might think, drawing drastic conclusions from NP is somewhat of a stretch. The problem is not just that the scenario envisaged is rather fanciful. Rather, it is that it seems incompletely specified. In particular, we are simply told that the Predictor is highly reliable. No explanation is given of how and why her predictions are so reliable. As J. L. Mackie pointed out in an excellent but somewhat neglected paper published in 1977, different explanations of the Predictor's reliability point to different choices in NP. Some explanations favor one-boxing and others, two-boxing. So, for example, if the agent is capable of some kind of backward causation, then one-boxing seems clearly to be the rational response.

In fact, Mackie makes a very strong claim, suggesting that NP is ill-formed.

The survey as a whole, therefore, shows that every possible case is somehow off-color. There is no conceivable kind of situation that satisfies at once the whole of what it is natural to take as the intended specification of the paradox. We simply cannot reconcile the requirements that the player should have, in a single game, a genuinely open choice, that there should

---

[1] See the next section for a fuller discussion of exactly how the expectation should be calculated here.
[2] See, e.g., Nozick 1969; Gibbard and Harper 1978; Skyrms 1980; Lewis 1981a. For dissenting views, see Horgan 1981; Eells 1981, 1982; Ahmed 2014a.

be no trickery or backward causation, and that the seer's complete predictive success should be inductively extrapolable. While the bare bones of the formulation of the paradox are conceivably satisfiable, what they are intended to suggest is not. The paradoxical situation, in its intended interpretation, is not merely of a kind that we are most unlikely to encounter; it is of a kind that simply cannot occur. (1977: 223)

Even someone sympathetic to Mackie's position, though, will think that he is overplaying his hand. How can we be so sure that he has canvassed all the available explanations of the Predictor's success? At the same time, though, nobody who is skeptical about NP's utility in settling disputes about the nature of rational decision-making is likely to think it profitable to think up increasingly baroque explanations of the extraordinary powers of a mysterious and possibly supernatural entity.

So, there is something of an impasse. One way to break it (perhaps the only way) is to turn attention away from the original NP and look instead at the real-life decision problems that have been claimed to have the same basic structure as NP. Plainly, if there are such real-life NPs, then they will include some sort of analog to the highly reliable Predictor. Crucially, moreover, there will be built into the decision problem (or at least, easily extractable from it) an explanation of why the Predictor-analog is so reliable. And so, if there are any real-life NPs, then Mackie's worry will have been answered.

The first step in thinking about whether there are real-life NPs is to give an abstract specification of the original NP to which real-life decision problems can be compared. Such an abstract specification is given in section 1. In section 2 I consider a representative example of the class of medical decision problems that have been claimed to exemplify the basic structure of NP. Section 3 considers a putative economic NP, while section 4 evaluates the frequently made claim that NP is really equivalent to the Prisoner's Dilemma (PD), which itself has a wide range of real-life exemplars.

1 The Structure of NP

What sort of structure would a decision problem have to satisfy in order to count as a real-life NP?

Any NP is specifiable in normal form in a 2 x 2 matrix, as represented in Table 2.

Table 2

	S_1	S_2
A	a_1	a_2
B	b_1	b_2

The decision-maker has two available courses of action (A and B) and there are two relevant possible states of the world (S_1 and S_2). This yields four possible payoffs (a_1, a_2, b_1, and b_2).

First, in order for a decision problem to count as an NP, there must be a candidate for a dominant action. This means that the payoff from one available action must be more highly valued than the payoff from the other, irrespective of which possible state of the world is actual. Without loss of generality, I will assume in the following that B dominates A, so that $b_1 > a_1$ and $b_2 > a_2$.[3]

Second, it is characteristic of NPs that the less highly ranked payoff in one state of the world should still be higher than the more highly ranked payoff in the other state of the world. So, again without loss of generality, I will assume that $a_1 > b_2$. Combining the first two conditions gives the following ranking of the payoffs:

$$b_1 > a_1 > b_2 > a_2.$$

The third condition has to hold in order for the candidate dominant action really to count as a dominant action. This is that both states of the world must be causally independent of the agent's choice of action. To see why this is important, suppose that performing action A has a high probability of bringing about state S_1 and a low probability of bringing about state S_2, while performing B has a high probability of bringing about S_2 and a low probability of bringing about S_1. Then, the decision-maker is really choosing between a_1 and b_2, and, since (by condition 2) we have that $a_1 > b_2$, action A is to be preferred.

The fourth condition satisfied by all Newcomb Problems is that there be a high, but non-causal, probabilistic dependence between A and S_1, on the one hand, and B and S_2, on the other. In the original Newcomb Problem, the high reliability of the Predictor means that the two conditional probabilities $p(A/S_1)$ and $p(B/S_2)$ are both close to 1. As Levi observed, this is perfectly

---

[3] Nozick has pointed out that the size of the intervals is important. Intuitions in favor of dominance seem to diminish when the gap between the B-outcomes and the A-outcomes is very small (imagine that there is only a penny in the transparent box, for example). See Nozick 1993: 44–5.

compatible with the converse probabilities $p(S_1/A)$ and $p(S_2/B)$ being significantly lower (Levi 1975). However, as standardly developed, in Jeffrey-Bolker decision theory, for example, calculations of expected utility employ the converse probabilities. What matters is the probability of the world being in a given state, conditional upon your acting in a certain way. So, if the converse probabilities are low, then expected utility calculations may not yield a different prescription from the dominance principle. Levi made heavy weather out of this, in my opinion. It is simple enough just to require that $p(S_1/A)$ and $p(S_2/B)$ are also close to 1, and so I will do so.[4] (Of course, Mackie's point still stands – we need a plausible story about *why* these conditional probabilities are close to 1).

Finally, the interest of NPs lies in the apparent conflict that they display between EDT and CDT, and so any candidate for a real-life NP will have to be such that EDT and CDT make conflicting recommendations. Again, without loss of generality, in the current framework EDT will have to recommend A in virtue of the high conditional probability $p(S_1/A)$, while CDT will have to recommend B by dominance.

In sum, then, to count as an NP a decision problem with a payoff matrix as in Table 2 must satisfy the following five conditions –

(1) $b_1 > a_1$ and $b_2 > a_2$
(2) $a_1 > b_2$
(3) Each of S_1 and S_2 must be causally independent of both A and B
(4) S_1 and S_2 must be probabilistically dependent upon A and B, respectively
(5) EDT and CDT must yield conflicting prescriptions, with EDT recommending A and CDT recommending B.

With this characterization in hand, we can consider candidates for real-life NPs.

2 Medical Newcomb Problems?

Perhaps the most frequently discussed examples of putative real-life NPs are medical in nature.[5] Actually, it is more accurate to describe them as quasi-medical, since they typically involve flights of fancy not widely shared within the medical community (e.g., that the correlation between smoking and lung cancer is due to their both having a common cause, the so-called smoking

---

[4] Cf. Horgan 1985: 224.
[5] For examples of so-called medical Newcomb Problems, see Nozick 1969; Skyrms 1980; Horgan 1981; Lewis 1981a; Eells 1982; Price 1986.

lesion). But still, the scenarios envisaged are perfectly comprehensible and it is not hard to imagine that the world might have turned out to be the way they describe. The medical NPs all seem to me to be variations on a single theme, and so I will just take one example and happily generalize from it. In order to give the smoking lesion some well-deserved rest, this section will focus on a structurally similar cholesterol example originally due to Brian Skyrms.[6]

Suppose that the well-documented correlation between hardening of the arteries and high cholesterol intake is due, not to the latter causing the former, but rather to the fact that people with hardened arteries and high cholesterol intake have a distinctive kind of lesion in the walls of their arteries. This lesion has two effects in those who have it. Their arteries harden and they increase their cholesterol intake. To repeat. This is not the way the world is, but it is certainly easier to make sense of the possibility of the cholesterol lesion than it is to make sense of the possibility of an almost infallible predictor.

Relative to this background story about the cholesterol lesion, imagine that you are faced with the following decision problem. You are deciding what to have for breakfast. In fact, following a suggestion from Ellery Eells, imagine that you are choosing not just for today, but for the rest of your life. The choice is a simple one. You can have Eggs, or you can have Fruit & Yogurt. There are two relevant states of the world. In one state you have the Lesion, while in the other you have No Lesion. You definitely prefer Eggs to Fruit & Yogurt, irrespective of whether you have the Lesion or not. But still, you would very much rather have Fruit & Yogurt in the No Lesion state of the world than Eggs in the Lesion state. So, putting in some representative numbers, your decision problem can be captured by the following matrix.

Table 3

	No Lesion	Lesion
Fruit & Yogurt	9	0
Eggs	10	1

This is a fanciful case, but still close enough to reality to count as a real-life NP – provided that it satisfies the five conditions identified earlier. So, we need to go through them and check. I have presented the decision matrix so that it is easily mapped onto our prototypical NP. We can take A = Fruit &

---

[6] See Skyrms 1980: 128–9.

Yogurt, B = Eggs, S_1 = No Lesion and S_2 = Lesion. This all gives us $a_1 = 9$, $b_1 = 10$, $a_2 = 0$ and $b_2 = 1$.

The first three conditions are obviously satisfied. We have $b_1 > a_1$ and $b_2 > a_2$. And we also have $a_1 > b_2$. So, the payoff structure is isomorphic to the original NP. Moreover, the way the scenario is described secures the causal independence of S_1 and S_2 from A and B. The whole point of stipulating (counterfactually) that there is a common cause of hardened arteries and high cholesterol intake is to ensure that there is no direct causal relation between your cholesterol intake and the state of your arteries.

What about (4) and the requirement that S_1 and S_2 be probabilistically dependent upon A and B, respectively? Here is a way of thinking about the case on which condition (4) does seem to be satisfied. The first thing we need is that there be a high probability of Eggs, conditional upon Lesion, so that $p(B/S_2)$ is close to 1. Let's say that this follows from the fact that the lesion causes high cholesterol intake. As observed earlier, though, calculations of expected utility standardly operate via the probability of the state of the world, conditional upon a given action being performed. So, what we need is a high probability of Lesion, conditional upon Eggs – i.e., that $p(S_2/B)$ be close to 1. It looks as though we can secure this with the additional assumption that having the lesion is the most likely explanation for choosing Eggs.

Two parallel assumptions seem to secure the probabilistic dependence of S_1 upon A (and vice versa). So, it looks as though we have condition (4).

This leaves only condition (5), which is that EDT and CDT must give conflicting prescriptions. Let's start with CDT. From the perspective of Causal Decision Theory, what matters is not the probability of a state of the world conditional upon an action, but rather the probability that the action will bring about that state of the world. Causal decision theorists typically think about the possible causal dependence of a state of the world upon an action in terms of the probability of the hypothesis that the action *would* bring about that state, if it were to be performed. Such causal hypotheses are standardly formulated as counterfactual conditionals and written as, for example, "A $\square\rightarrow S_1$." So, for the causal decision theorist, what matters is the probability of the relevant counterfactuals stating the potential states of the world that a given action might bring about, were it to be performed. In many cases, the probability of such a counterfactual will coincide with the probability of the state of the world conditional upon the action, but in NPs they are supposed to diverge. And where the probability of the counterfactual diverges from the associated conditional probability, causal decision theorists hold the former to be more fundamental.

So, in cases where a state of the world is causally independent of an action, then the probability of the associated counterfactual is equal to the *unconditional* probability of the state of the world occurring – since the action can do nothing to bring that state of the world about (or to prevent it from occurring). In the case under consideration, therefore, we have the following identities holding in virtue of the causal independence of S_1 and S_2 from A and B, respectively:

$$p(A \square \rightarrow S_1) = p(S_1), \tag{i}$$

$$p(B \square \rightarrow S_2) = p(S_2). \tag{ii}$$

These identities mean that dominance reasoning can be applied. So, CDT prescribes choosing Eggs. But what about the other half of condition (5), which is that EDT should make the conflicting prescription of Fruit & Yogurt?

Certainly, if the conditional probabilities are as suggested above, then EDT seems clearly to prescribe Fruit & Yogurt. Suppose, let's say, that

$$p(S_1/A) = p(S_2/B) = 0.9, \tag{iii}$$

$$p(S_2/A) = p(S_1/B) = 0.1. \tag{iv}$$

Then the expected utility of Eggs = (0.1 × 10) + (0.9 × 1) = 1.9, while the expected utility of Fruit & Yogurt = (0.9 × 9) + (0.1 × 0) = 8.1. Fruit & Yogurt wins.

However, a number of authors have objected to this way of thinking about the conditional probabilities. According to what has become known as the Tickle Defense, analogs to (i) and (ii) hold within EDT in exactly the same way that they do within CDT, even though EDT places no requirement of causal dependence upon conditional probability.[7] I will give an informal characterization of the Tickle Defense first, followed by a more formal version adapted from its originator, Ellery Eells.

The basic premise behind the Tickle Defense (as it might be applied in this case) is that the lesion operates through its effects on the decision-maker's psychology. We are told that the lesion causes those who have it to increase their cholesterol intake. And, since we are discussing decision principles for rational choice, it is a background assumption for the debate that, in this particular case and others like it, it is up to the decision-maker to decide

---

[7] See Ahmed 2014a; Eells 1982; Horgan 1981; Horwich 1987; Jeffrey 1983.

whether or not to take the high-cholesterol option of Eggs. As a decision-maker, in other words, you are making a genuine decision, where a genuine decision is one in which both of the available options are open to you. You are not in the grip of forces beyond your control (as a drug addict might be, for example), nor is your choice fixed by autonomic processes (in the way that your breathing is). So, it really is open to you to choose either Eggs or Fruit & Yogurt.

But how does a rational decision-maker decide whether to choose Eggs or Fruit & Yogurt? In the general decision-theoretic framework within which we are operating, rational decision-makers base their decisions on their utilities and the probabilities that they assign to different possible states of the world – which is really a way of regimenting our everyday talk about beliefs and desires.[8] So, we should assume that the only way the lesion can affect a decision-maker's choices is through affecting their utility and/or probability assignments. Without these psychological intermediaries, it is not clear that we have a decision problem at all.

But now we assume a reasonably self-aware decision-maker, who is well aware of the twofold effect of having the cholesterol lesion. Such a decision-maker knows that, if he has a desire for Eggs (manifested in his utility assignments), then this is very good evidence that he has the lesion, and hence is highly likely to have hardened arteries. So, he knows perfectly well that having the desire for Eggs would be an unwelcome development. More-over, we can suppose that our decision-maker has sufficient introspective acuity to know whether or not he has the desire for Eggs. This is a perfectly reasonable thing to suppose in the present context, since all it requires is that the decision-maker be familiar with the relevant payoffs for each of the two possible actions in each possible state of the world – or, in other words, that he be familiar with the payoff table that defines the decision problem he confronts.[9]

This is the crux of the Tickle Defense, which holds that since the decision-maker already has all this information, knowing how he will actually choose provides no new information about whether or not he has the Lesion (and hence about the condition of his arteries). If you have sufficient knowledge of your own mental states and how they will lead you to act, then you already have all the information that you are going to get about the presence/absence of the Lesion and the condition of your arteries. Knowing how you actually

---

[8] For further discussion, see Bermúdez 2009; Lewis 1994a; Okasha 2015; Pettit 1991.
[9] See note 11 below.

choose adds no new information. But then there is no difference between the conditional probability of the Lesion, given your choice, and the unconditional probability of the Lesion. So, EDT ends up with an equivalence directly parallel to the identities given earlier in (i) and (ii). For that reason, EDT will prescribe taking the dominant strategy just as CDT does. The conflict evaporates and condition (5) fails to hold.

So, that's the general idea. Now let's look at Eells's original version of the Tickle Defense (transposing it into the current framework).[10] Eells makes an important distinction between two different, but superficially similar, conditional probability judgments. The first is a general judgment about an entire population. Based on your medical 'knowledge' of cholesterol lesions, you might be very confident that the following two inequalities hold:

$$p(\text{Lesion}/\text{Eggs}) > p(\text{Lesion}/\text{Fruit \& Yogurt}), \qquad (v)$$

$$p(\text{No Lesion}/\text{Fruit \& Yogurt}) > p(\text{No Lesion}/\text{Eggs}). \qquad (vi)$$

Eells glosses (v) and (vi) as being, in essence, claims about how artery condition and diet are correlated across the population as a whole. If you come across two random individuals and discover that the first typically chooses Eggs, while the second typically chooses Fruit & Yogurt, then you would reasonably take the first person to be more likely to have the Lesion. But he is very careful to distinguish (v) and (vi) from the conditional probabilities that matter to you in your decision whether to have Eggs or Fruit & Yogurt. The conditional probabilities that matter to you for *your* deliberation are probabilities about the condition of *your* arteries, conditional upon *your* choice. Let's write these as follows:

$$p(\text{Lesion}_{\text{ME}}/\text{Eggs}_{\text{ME}}) > p(\text{Lesion}_{\text{ME}}/\text{Fruit \& Yogurt}_{\text{ME}}), \qquad (vii)$$

$$p(\text{No Lesion}_{\text{ME}}/\text{Fruit \& Yogurt}_{\text{ME}}) > p(\text{No Lesion}_{\text{ME}}/\text{Eggs}_{\text{ME}}). \qquad (viii)$$

Eells's fundamental point is that (v) does not entail (vii) and (vi) does not entail (viii). Why not? Because you know things about yourself that you do not know about other people.

Let "φ" denote the set of propositions about your own probability and utility assignments that will fix the expected utility of your two available actions, and

---

[10] Eells's original arguments are in Eells 1982. I am working primarily from Eells 1985, which is a distillation of ch. 3–6 from that book, together with material from Eells 1984.

let "Rφ" denote the proposition that you know every proposition in φ. Unless Rφ holds, you will be unable to fix expected utilities for either available action, and so it is reasonable to assume that you know this and therefore assign a probability 1 to Rφ:[11]

$$p(R\varphi) = 1. \tag{ix}$$

Now, let "MAX (Eggs)" denote your determining that Eggs is the action that maximizes expected utility, and similarly for "MAX (Fruit & Yogurt)." Since you know that you are a rational decision-maker (and, ex hypothesi, your criterion for rational decision-making is maximizing expected utility, and you know this), you are confident that you will choose Eggs if and only if Eggs is the action that maximizes expected utility, which gives:

$$p\big(\text{MAX } (\text{Eggs}_{ME}) \Leftrightarrow \text{Eggs}_{ME}\big) = 1. \tag{x}$$

Both (ix) and (x) seem to be reasonable idealizations.

The last assumption needed for Eells's version of the Tickle Defense is this:

$$\begin{aligned} p\big(\text{MAX } (\text{Eggs}_{ME})/R\varphi \text{ & Lesion}_{ME}\big) \\ = p\big(\text{MAX } (\text{Eggs}_{ME})/R\varphi \text{ & No Lesion}_{ME}\big). \end{aligned} \tag{xi}$$

What (xi) says is that whether or not you have the lesion is evidentially irrelevant to how you determine whether Eggs maximizes expected utility, given the probability and utility assignments that you happen to have. Why is this? Because whether or not Eggs maximizes utility is fully determined by your probability and utility assignments, and those assignments are what they are. The presence or absence of the lesion is irrelevant to whether or not Eggs maximizes expected utility.

With all these pieces in place it is straightforward to show that (vii) does not hold, but that instead we have a version of the identities in (i). First, we can combine (x) and (xi) to give:

---

[11] A few more words on these assumptions. To be in a (putative) Newcomb Problem is not just to be in a situation where a third-person observer might observe that CDT and EDT give conflicting recommendations. Newcomb's Problem is supposed to be a first-person dilemma – a situation where the conflict between CDT and EDT is manifest to the decision-maker. For that to be the case, however, the decision-maker must herself be able to reason her way to each of the conflicting recommendations, which in turn requires that she know her probability and utility assignments and know that she is a maximizer of expected utility. So, the assumptions in the text are really necessary idealizations.

$$p(\text{Eggs}_{\text{ME}}/\text{R}\varphi \And \text{Lesion}_{\text{ME}}) = p(\text{Eggs}_{\text{ME}}/\text{R}\varphi \And \text{No Lesion}_{\text{ME}}).$$

(xii)

So, the probability that I will choose Eggs conditional upon my probability/ utility assignments and my having the Lesion is the same as the probability of my choosing Eggs conditional upon my probability/utility assignments and my not having the Lesion. We can now plug in (ix), which says that $p(\text{R}\varphi) = 1$, to simplify to:

$$p(\text{Eggs}_{\text{ME}}/\text{Lesion}_{\text{ME}}) = p(\text{Eggs}_{\text{ME}}/\text{No Lesion}_{\text{ME}}).$$

(xiii)

Parallel reasoning gives:

$$p(\text{Fruit and Yogurt}_{\text{ME}}/\text{Lesion}_{\text{ME}}) \\ = p(\text{Fruit and Yogurt}_{\text{ME}}/\text{No Lesion}_{\text{ME}}).$$

(xiv)

But, (xiii) and (xiv) together yield:

$$p(\text{Lesion}_{\text{ME}}/\text{Eggs}_{\text{ME}}) = p(\text{Lesion}_{\text{ME}}/\text{Fruit} \And \text{Yogurt}_{\text{ME}})$$

(xv)

and

$$p(\text{No Lesion}_{\text{ME}}/\text{Eggs}_{\text{ME}}) = p(\text{No Lesion}_{\text{ME}}/\text{Fruit} \And \text{Yogurt}_{\text{ME}}).$$

(xvi)

Or in other words, $p(S_1/A) = p(S_1/B) = p(S_1)$, and $p(S_2/A) = p(S_2/B) = p(S_2)$.

If this reasoning is sound, then condition (4) fails to hold, which means that (5) fails and the cholesterol lesion case fails to count as a genuine NP.

One of the advantages of looking at Eells's version of the case is that it makes explicit how the Tickle Defense rests upon a first person/third person asymmetry. This assumption has been challenged by Frank Jackson and Robert Pargetter, who make the following general observation about decision theory:

Decision theories deal with what it is right for me to do by my own lights, thus it is my (the agent's) subjective probability function that is used in calculating expected utilities. But, as well as what is right by my own lights, there is what is right by another's lights. As well as asking what is right (subjectively so, but we will take the qualification as understood from now on) for X to do by X's lights, we can ask what it is right for X to do by Y's lights... What we have in mind is not assessing the options available to another in terms of his value and probability functions, but in terms of his value and *our* probability function. (1983: 217)

Fleshing out their example a little more, I might have good reason to think that my aunt's probability function, while being internally consistent, is nonetheless irrational in various ways. So, I might want to reason on her behalf, but substituting my own probability function for hers. In that sort of situation, they claim, the conflict between CDT and EDT reappears, but there is no prospect of applying the Tickle Defense (since I would not be able to apply the same information about X as X has about X or I have about myself). In more detail:

> Suppose *I* want the answer to whether my aunt in Scotland should smoke. In the smoker's fantasy case, my answer is clearly that she should... But EDT gives as my answer the wrong one that she should not smoke, for my probability of her dying given she smokes is greater than that given she does not... There is no mileage to be gained on behalf of EDT out of the Tickle Defense. I know nothing of my aunt's cravings or 'tickles,' no matter how transparent her mind is to herself. (ibid.: 217–8)[12]

In a Postscript added to a reprint of the paper, they explicitly apply this objection to Eells's version of the Tickle Defense, observing that Eells's common cause cannot "affect an outside observer's deliberations concerning what, by his lights, the agent should do" (ibid.: 218).

I find this objection ingenious but unconvincing. Let's go back to Eells's argument. The relevant premise hinges on what he terms φ, which is the set of propositions about the agent's own probability and utility assignments that will fix the expected utility of the available actions. Plainly, in the case that Jackson and Pargetter consider, φ is the set of propositions that gives all the information required to fix my probability assignments and my aunt's utility assignments. If I am to solve the decision problem at all, then I must know every proposition in φ. I may not know my aunt's cravings and tickles, but for the objection even to get off the ground, then I must know my aunt's utility function to exactly the same extent that I must know my own – i.e., to the extent required to fix the expectation of her two available actions. Hence, we have $R\varphi$. Moreover, it seems a reasonable assumption that I be completely confident that $R\varphi$ holds. So, from my perspective $p(R\varphi) = 1$. It is fair to

---

[12] Note that Jackson and Pargetter use "CDT" to abbreviate "conditional decision theory," which is their label for what is more typically termed Evidential Decision Theory. I have modified their terminology to avoid confusion.

assume, also, that, when reasoning on my aunt's behalf, I attribute to her complete confidence in the belief that any (utility-maximizing) rational agent will have, namely, that she will perform the action with the highest expected utility. But, as we have seen, that is all that is required for Eells's argument to run. The most plausible way of looking at Jackson and Pargetter's suggestion is that I am correcting for problems in my aunt's probability function (which I might have good grounds to believe to be irrational). But the probabilistic judgments required to run the Tickle Defense are ones that will plausibly be constant across any rational agent, so they will hold when I am placing myself in my aunt's shoes, or she in mine. And so, the Tickle Defense can be applied here too, despite the Jackson and Pargetter objection.

I conclude that so-called medical NPs fail to count as NPs. My confidence in generalizing from Skyrms's cholesterol case (and the brief discussion of smoking lesion) has two grounds. The first is that all medical NPs adopt some sort of common cause hypothesis to generate an alleged conflict between EDT and CDT, and the Tickle Defense can be applied whenever there is a common cause structure. And the second is that (as should be clear from the discussion of Jackson and Pargetter's objection), the assumptions required to run the argument are really very weak and would be satisfied by any rational agent.

3 Economic Newcomb Problems?

The suggestion that NP-type situations can be found in macroeconomics was first put forward, to the best of my knowledge, in a 1982 paper in the *Southern Economic Journal* by Frydman, O'Driscoll and Schotter, and subsequently developed by John Broome (who at the time was still officially an economist) in a short piece in *Analysis* in 1989. I will follow Broome's presentation (1989, 1990a).

Consider the decision-making unit in a central bank (say, the Open Market Committee of the Federal Reserve in the United States, or the Bank of England's Monetary Policy Committee). The committee is trying to decide whether or not to expand the money supply (which it can do, for example, by lowering the amount of funds that banks must hold in reserve against deposits, thereby allowing banks to lend out more money). There are definite benefits to increasing the money supply. Standard theory says that increasing the money supply will increase employment, for example.

But the committee faces a dilemma, because there is a strong probabilistic interdependence between the money supply and public expectations of the money supply. Generally speaking, the expectations that the public has are good predictors of what the central bank committee will decide to do. But if the central bank increases the money supply, and the public has predicted that that is what it will do, then the result will be inflation. If the central bank keeps the money supply constant, and the public has anticipated this, then the status quo is the most likely outcome. But if the central bank surprises the public by keeping the money supply constant, then the result will be recession. The best outcome (increased employment) will occur only if the central bank surprises the public by increasing the money supply when it was expected to keep it constant.

Here is Broome's description of the situation:

> If the government expands the money supply, the people will probably have predicted that, so the result will be inflation. If it does not expand it, they will probably have predicted that too, so the result will be no change. The Bolker-Jeffrey theory [i.e., EDT], then, will assign a higher expected utility to not expanding. It suggests that this is the right thing to do. Dominance reasoning, however, shows that the right thing is to expand. That, at any rate, is the conclusion of most authors who have considered this 'time-inconsistency problem'. The government's dilemma has exactly the form of the 'Newcomb Problem', which first led to the interest in causal decision theory. (1990a: 104)[13]

And here is a 2 × 2 matrix representing the decision problem in a way that lines up with our other example (I have added numbers consistent with the above discussion).

Table 4

	No expansion expected	Expansion expected
No expansion	*Status quo* (9)	*Recession* (0)
Expand	*Increased employment* (10)	*Inflation* (1)

As before, it is not hard to see that the first three of our conditions are satisfied, if we take A = No Expansion. B = Expansion. S_1 = No expansion expected. S_2 = Expansion expected and so on. The outcomes are ranked

---

[13] Note that Broome was writing in the UK before the Bank of England was granted operational independence over monetary policy after the Labor Party victory in the 1997 general election. Hence his reference to the government expanding the money supply.

$b_1 > a_1 > b_2 > a_2$, thus satisfying conditions (1) and (2). Moreover, the public's expectations cannot depend causally upon what the central bank actually chooses to do, not least because they precede the bank's actions. This yields (3).

Given (3) and the payoff table, there is a clear dominance argument in favor of Expand. So, we will have an economic NP if and only if (4) holds, that is, if there is a strong probabilistic interdependence between what the central bank decides to do and what the public expects. Broome and others assume that $p(S_1/A)$ and $p(S_2/B)$ are high. In particular, they assume the inequalities $p(S_1/A) > p(S_1/B)$ and $p(S_2/B) > p(S_2/A)$. Actually, though, this is unwarranted, as is shown by a simple and elegant argument originally due to Arif Ahmed.[14]

The starting-point is that public expectations about money supply are based on evidence (about what has happened in comparable macroeconomic situations in the past, the guidance that the central bank has provided, current economic indicators and so forth). Let "ψ" denote the set of propositions that fully characterizes the public's evidence-base. As before, let "φ" denote the set of propositions that fully characterizes the central bank's evidence-base. For any minimally competent central bank, we have that $\psi \subseteq \varphi$. Moreover, we can assume both that $R\varphi$ is the case (where "$R\varphi$" says that all the propositions in φ are known to the central bank), and that the central bank knows that $R\varphi$ is the case – i.e., that $p(R\varphi) = 1$.

Now consider the following conditional probabilities:

$$p(\text{Expansion Expected}/\psi \ \& \ \text{Expand}), \qquad \text{(xvii)}$$

$$p(\text{Expansion Expected}/\psi \ \& \ \text{Not Expand}). \qquad \text{(xviii)}$$

Ahmed points out that these two probabilities must be the same (from the perspective of the central bank), because whether Expansion Expected holds is fully determined by ψ. A rational central bank must hold its own actions to be evidentially irrelevant to how public expectations are formed, because the public forms its expectations on the basis of ψ. Hence:

$$p(\text{Expansion Expected}/\psi \ \& \ \text{Expand})$$
$$= p((\text{Expansion Expected}/\psi \ \& \ \text{Not Expand})$$
$$= (\text{Expansion Expected}/\psi). \qquad \text{(xix)}$$

---

[14] See Ahmed 2014a: §4.4. As he puts it, the argument is not a straightforward application of the Tickle Defense, but it is nonetheless ticklish.

Reasonably, if we have $p(R\varphi) = 1$ and $\psi \subseteq \varphi$, then we have $p(R\psi) = 1$. This allows us to simplify (xvii) to:

$$p(\text{Expansion Expected}/\psi) = p(\text{Expansion Expected}). \qquad \text{(xx)}$$

Parallel reasoning gives:

$$p(\text{No Expansion Expected}/\psi \& \text{Expand})$$
$$= p(\text{No Expansion Expected}/\psi \& \text{Not Expand})$$
$$= p(\text{No Expansion Expected}/\psi) = p(\text{No Expansion Expected}). \qquad \text{(xxi)}$$

So, we have the same identities as in the medical NP: $p(S_1/A) = p(S_1/B) = p(S_1)$ and $p(S_2/B) = p(S_2/A) = p(S_2)$. Condition (4) fails to hold, and so there is no conflict between EDT and CDT, which means that condition (5) fails to hold. Both EDT and CDT recommend the dominant action of Expansion.

The economic candidate for a real-life NP fails to qualify in exactly the same way as the medical examples, albeit for slightly different reasons.

4 Is the Prisoner's Dilemma a Newcomb Problem?

The last candidate I will consider for a real-life NP is the famous and much-discussed Prisoner's Dilemma (henceforth: PD). On the face of it, showing that the PD is a real-life NP would be enough to allay all concerns about NP being a genuine and well-specified decision problem, given how many different types of social interaction can profitably be modeled as PDs. The most developed argument for taking NP and PD to be notational variants of a single problem was made by David Lewis in a short paper in 1979, which ends with the assertion that 'Prisoners' Dilemmas are deplorably common in real life. They are the most down-to-earth versions of Newcomb's Problem now available.' His conclusions have been widely, indeed almost universally, accepted.[15]

Lewis's assimilation of NP and PD begins by making some inessential changes to the payoff table for PD, which he formulates like this (with A's payoff first in each case):

---

[15] Mild dissent can be found in Sobel 1985a, which offers the gentle qualification that some logically possible PDs are not NPs. For further discussion of the relation between NP and PD, see Pettit 1988; Hurley 1991, 1994. I have criticized Lewis's argument in Bermúdez 2013, 2015a, 2015b – but see Walker 2015.

Table 5

	B cooperates	B does not cooperate
A cooperates	$1,000,000, $1,000,000	$0, $1,001,000
A does not cooperate	$1,001,000, $0	$1,000, $1,000

In essence, not cooperating is rewarded with $1,000 whatever happens, while each player receives $1,000,000 only if the other player cooperates (by declining their $1,000). From A's point of view, therefore, this version of the PD has the following basic structure:

(1) I am offered $1,000 – take it or leave it.
(2) I may be given an additional $1,000,000, but whether or not this happens is causally independent of the choice I make.
(3) I will receive $1,000,000 if and only if you do not take your $1,000.

Here for comparison is how NP looks to the Newcomb chooser:

(1) I am offered $1,000 – take it or leave it.
(2) I may be given an additional $1,000,000, but whether or not this happens is causally independent of the choice I make.
(3^*) I will receive $1,000,000 if and only if it is predicted that I do not take my $1,000.

Lewis argues that in the right circumstances (3) and (3^*) turn out to be equivalent. When the two players in PD are sufficiently similar, and believed to be sufficiently similar, each player can take the other's choice as potentially predictive of her own choice. So, from A's point of view, B's not taking his $1,000 predicts A's not taking her $1,000 – and vice versa.[16] Lewis's claim, therefore, is that each player in a PD so construed effectively confronts a version of NP, so that the PD is really just a pair of NPs.[17]

My basic objection to Lewis's argument is that the variations that he makes to the PD in order to make it look like an NP effectively transform the game so that it is no longer a PD. One reason to be suspicious of this strategy right from the start is that it is plainly impossible to present the PD in the canonical form set out in section 1. The standard presentation of the PD in normal form certainly takes the form of a 2 × 2 matrix, but the outcomes are joint, not

---

[16] In the following, I adopt the convention by which player A is female and player B is male.
[17] In the following, when I talk about the claim that the PD is an NP, what I mean, of course, is the more circumscribed thesis that each player confronts a form of NP, so that the entire PD interaction comes out as a pair of NPs.

individual. In effect, what I will be arguing in the following is that this impossibility reflects a fundamental difference between the two problems. It is not an accident of notation.

Here is a warm-up observation. Suppose, first, that I have a very high degree of confidence that the other player is similar to me, so that in effect I take her to be a replica of me. This means that I must have a very low degree of confidence in the genuine possibility of my taking my $1,000 while the other player does not take her $1,000 – and similarly a very low degree of confidence in the genuine possibility of my not taking my $1,000 while the other player takes her $1,000. At the limit, where I believe that the other person is a perfect replica of me, I am committed to thinking that the two scenarios just envisaged are impossible.

In earlier papers, I observed that player A cannot simultaneously believe all of those things while still thinking that she is in a PD, since in effect she is thinking that two of the four available scenarios in the PD's payoff matrix are impossible.[18] What player A is committed to believing is that her only two live alternatives are the upper left and bottom right scenarios in the matrix – the scenarios where both players cooperate or where they both fail to cooperate. This is a completely different decision problem, and so, I argued, Lewis's position is unstable. The reasoning that leads him to conclude that the PD is a notational variant of NP effectively changes the game so that it is not a PD.

This reasoning only applies, though, to what one might think of as canonical NPs – that is to say, NPs where $p(S_1/A)$ and $p(S_2/B)$ are at or close to 1. This is how NP is almost always presented, but as Arif Ahmed has observed (and as Lewis himself points it out in his original paper), generating a conflict between EDT and CDT does not require that this hold. All that is required is that $p(S_1/A)$ and $p(S_2/B)$ be slightly more than 0.5 (> 0.5005, to be precise).[19]

A more fundamental objection is that the existence of any kind of probabilistic dependence between player A's choice and player B's choice means that this redefined PD no longer qualifies as an instance of strategic choice – a game in the technical sense. Here is the classic characterization of a game, from Luce and Raiffa's famous textbook:

> There are *n* players each of whom is required to make one choice from a well-defined set of possible choices, and these choices are made without any knowledge as to the choices of the other players... Given the choices of each of the players, there is a resulting outcome, which is appraised by each of the players according to his own peculiar tastes and preferences. The

[18] See Bermúdez 2013, 2015a, 2015b.   [19] Ahmed 2014a: 113–4.

problem for each player is: What choice should be made in order that his partial influence over the outcome benefits him the most? He is to assume that each of the other players is similarly motivated. (1957: 5–6)

The key points here are, first, that games are strategic interactions where each player has only a partial influence on the outcome; second, that each player has a preference order over the possible outcomes and, third, that in games each player chooses in ignorance of the other players' choices.

The last point is telling. The decision problem that Lewis describes is not one where each player chooses in ignorance of what the other player is choosing, because of the probabilistic dependence between what I do and what the other player does. Admittedly, in games of complete information, each player knows the structure of the game and so knows which possible actions the other players have available to them. For that reason, it might be thought that common knowledge of rationality implies knowledge that the Nash equilibrium will be played in games where there is a unique pure strategy equilibrium. Actually, though, this is debatable. It is not irrational to think that the non-Nash equilibrium outcomes are possible. It is not inconsistent to think that even an agent known to be perfectly rational can have a lapse. An agent who makes the rational choice is still perfectly capable of acting otherwise. The words might come out wrong, for example, or there could be some other type of 'tremble' that prevents a decision being implemented as intended.[20] So, at best, common knowledge of rationality is knowledge of how the players reason and choose. It does not yield certainty that the rational outcome will come to pass.

In any event, the point I want to extract from Luce and Raiffa's third condition is different. To appreciate it, observe first that it can often make sense to think of each player as assigning probabilities to the other players acting in certain ways. This is how mixed strategies are sometimes interpreted

---

[20] It is often argued that we need to allow the possibility of non-equilibrium behavior in order to establish that a particular outcome is the equilibrium outcome. The standard argument showing that (Defect, Defect) is the Nash equilibrium in PD depends upon counterfactuals such as: 'If player 1 were to Cooperate, then the best response for player 2 would be to Defect.' But the antecedents of all such counterfactuals would necessarily be false if out-of-equilibrium behavior were impossible. This so-called paradox of backward induction is one motivation for introducing the idea of a 'trembling hand,' or the possibility of 'a slip 'twixt cup and lip,' which allows any non-equilibrium path (in an extensive form game) to occur with non-zero-probability. See Selten 1975 for the original development of the idea of a trembling hand equilibrium, and Stalnaker 1996 for a more general discussion of counterfactuals in game theory.

(following, for example, Aumann 1987). But that is not inconsistent, I believe, with Luce and Raiffa's claim that game-theoretic choices are made 'without any knowledge as to the choices of the other players.' There is a fundamental distinction between having a probability distribution over the other player's possible actions, on the one hand, and knowing that what the other person will do is a (probabilistic) function of what you will do. In a typical game, each player tries to formulate their best response to what the other players might do. But the other players are independent variables, not dependent ones. As Yav Shoham and Kevin Leyton-Brown put it in a recent textbook, 'Game theory is the mathematical study of interaction among independent, self-interested agents.'[21] Or, to put it another way, the probability distribution that a player has over the other player's pure strategies is made up of unconditional probabilities, not probabilities conditional upon the player herself acting in a certain way. That is the real difference between the decision problem that Lewis proposes as a PD and the genuine PD. In the genuine PD, as in game theory more generally, what other players do is probabilistically independent of what I do, and the probabilities that I assign to how other players will choose must reflect this.[22]

In sum, there is a clear dilemma for any defender of Lewis's strategy. Lewis is trying to describe the PD so that it comes out looking like an NP. The first horn of the dilemma is the problem that we have just considered. In effect, Lewis's strategy runs the risk of doing violence to the PD by transforming it from a problem of strategic choice into a problem of parametric choice.[23] The only way of avoiding this risk is to redescribe the putative NP-version of the PD in such a way that it really is a strategic choice problem. I can't see how that can be done, but I assume for the sake of argument that it can be done. A defender of Lewis's position will now be caught on the second horn of the dilemma. If the redescribed PD really is a strategic choice problem, then it

---

[21] Shoham and Leyton-Brown 2009: 47.

[22] These comments should be taken to apply to noncooperative game theory (of which, of course, the PD is a paradigm example). The situation is more complicated in cooperative or coalitional game theory. For an introduction to key concepts in coalitional game theory, see Shoham and Leyton-Brown 2009, ch. 12.

[23] An agent in a parametric choice situation seeks to maximize expected utility relative to parameters that are set by the environment, and the rationality of her choice depends only on her utility and probability assignments, within the context set by the parameters. In strategic choices, in contrast, there is at least one other player, and as many dimensions of variation as there are players. The rationality of the agent's choices depends not just on her preferences and probability assignments, but also on the choices independently made by the other players – with the rationality of those choices also being partly determined by the agent's choices.

can't be an NP, because NP plainly fails to count as a strategic choice problem. This is so for three reasons:

- The Predictor is a predictor, and so, even if he is only accurate slightly more than 50.05% of the time, he cannot be described as acting 'without any knowledge as to the choices of the other players.'
- The Predictor does not strictly speaking choose at all. What the Predictor does is a (stochastic) function of what the player does.
- The Predictor does not have a preference ordering defined over the possible outcomes and so cannot qualify as a player.

Of these three, the most important is the second, which effectively reiterates the earlier point about conditional vs. unconditional probabilities from the opposite direction.[24] What defines a strategic interaction is that each player's actions are independent of the actions of the other players. As a player, I can certainly assign probabilities to what other players will do, but not in any way does that suggest that their choices will be probabilistically dependent upon what I choose. For that reason, a Newcomb-type scenario simply cannot occur within the standard framework of game theory.

This is important because it rules out the possibility of arguing, as Lewis does, that Newcomb Problems are 'deplorably common in real life.' There is considerable and familiar evidence for thinking that many social interactions instantiate the general structure of the PD. If it were indeed the case that each

---

[24] Arif Ahmed (2014a: 117–19) has suggested that voting in large elections can count both as a multi-agent PD and as an NP, provided that the following three conditions hold: (a) nonvoting dominates voting, because of the inconvenience of voting; (b) any voter in a large election should be to all intents and purposes certain that their vote will not make a difference; (c) voters often take their vote to be diagnostic. As Ahmed observes, many people vote, despite (a) and (b). He proposes (c) as a possible explanation. If so, there is potentially an argument to be made that conditions (1) through (5) are satisfied. Giving this proposal the attention it deserves would take us too far afield. I hope to address it in later work. However, here are two comments. The voting scenario does not have a predictor, and the other voters certainly have preferences over the possible outcomes. So, the first and third reasons for not taking NP to be a strategic choice problem do not apply. But the second does. To take your vote to be diagnostic is incompatible with taking other voters to be independent. And for that reason, the voting case cannot be a PD, in my view. But still, you might think that leaves open the possibility that it counts as a real-life NP. I wonder, though, about the payoff table. Attitudes to voting are very complicated, bringing into play all sorts of loyalties, obligations and perhaps what Nozick has called symbolic utility. So, I wonder about the assumption that nonvoting dominates voting. I am also not convinced that generally speaking people do take their votes to be diagnostic. Ahmed cites evidence that students planning to vote Yes in a 1992 Canadian referendum estimated that a higher proportion of the electorate would vote Yes than students planning to vote No. But that does not show that they take their vote to be diagnostic. They could, after all, be planning to vote Yes because they hope to 'surf the wave,' as it were. This is another case where we need more detail about the backstory.

player in a PD effectively confronts a version of NP, then NP-enthusiasts would really have an embarrassment of riches when it comes to finding real-life versions of NP. But sadly they do not, even when Lewis's assumptions are granted. NP and the PD are fundamentally different beasts.

5 Conclusion

I began with Mackie's complaint that NP is ill-defined. In particular, he argues, the appearance of conflict between EDT and CDT arises only because no backstory is provided as to how and why the crucial conditional probabilities have the values that they are claimed to have. It is certainly possible to spell out the backstory in sufficient detail to yield a well-defined decision problem, but doing so makes EDT and CDT agree. Mackie claims that there is no way of spelling out the backstory that will preserve the apparent conflict. The appearance of conflict is an artifact of a poorly specified decision problem.

I am sympathetic to this general claim, but unfortunately Mackie provides no reason to believe that he has considered all possible backstories for NP. And how could he? My more modest aim in this chapter has been to consider decision problems that have been claimed to be real-life NPs, counting on reality (or some close analog thereof) to do the job of fixing the backstory with sufficient clarity. The results have been disappointing for NP-enthusiasts, however. Relative to the abstract specification of an NP laid out in section 1, we can find real-life NPs neither in (fantasy) medicine nor in macroeconomics. The scenarios put forward as exemplifying NP certainly have the backstory spelled out in sufficient detail, but the conditional probabilities do not come out the way they would have to in order to generate a genuine conflict between EDT and CDT. Nor does the Prisoner's Dilemma fare any better as a real-life NP. It is the wrong kind of decision problem – a problem of strategic choice, rather than parametric choice. Richard Jeffrey famously described NP as a 'Prisoner's Dilemma for space cadets.' Unfortunately, he was only half right.

Instead of a conclusion, then, I end with a challenge to those who think that NP really does force a rethinking of evidential approaches to decision theory. Please provide a fully described decision problem in which conditions (1) through (5) hold, and where we have a convincing explanation of how and why a rational decision-maker can have the conditional probabilities required for those conditions to be met.

2   Newcomb's Problem, Rationality, and Restraint

Chrisoula Andreou

1 Introduction

There is a great deal of philosophical debate concerning whether it can be rational for an agent faced with a finite set of options to show restraint of the following sort: The agent incurs a cost or passes up a benefit even though she knows that doing so will bring about an outcome that she disprefers to another outcome that she could bring about instead. I will henceforth refer to this sort of restraint as *standard-form restraint*, since it is the form of restraint that is standardly at issue in philosophical debates about whether rationality sometimes calls for restraint.

According to one familiar line of thought, such restraint is irrational because rationality is nothing but instrumental rationality, and instrumental rationality prohibits forfeitures of the kind at issue. One might object to this line of thought based on the idea that rationality includes categorical imperatives – which commend certain objectives as in and of themselves worthy – and can thus require that an agent incur certain costs or pass up certain benefits even if doing so will bring about an outcome that she disprefers to another outcome that she could bring about instead. To this defense of standard-form restraint, it might be replied that if rationality includes categorical imperatives, then a rational agent's ranking of the options she faces will take these objectives into account, and so a rational agent's preferences will be shaped by these imperatives.

Even if this reply is accepted, there remains an altogether different way of defending the idea that it can be rational to show standard-form restraint. According to the alternative stance in question, such restraint can be rational even when nothing other than an agent's preferences are at stake and, indeed, even if there are no categorical imperatives. This may seem like an extremely odd position. For, insofar as all that matters is serving one's preferences well, how can it make sense to incur a cost or pass up a benefit even though one knows that doing so will bring about an outcome that one disprefers to another outcome that one could bring about instead? This is where a variety of intriguing puzzles concerning rationality and restraint enter the debate. The aim of this chapter is to venture into this terrain via a discussion of Newcomb's problem. Reviewing and building on existing work, I relate the problem to the toxin puzzle and the puzzle of the self-torturer, both of which are sometimes put forward as cases supporting the idea that, even given an instrumental conception of rationality, according to which rationality is about serving the agent's preferences, rationality can require standard-form restraint. Along the way, I discuss two variations of Newcomb's problem that have appeared in the literature and that are relevant in relation to at least some potential resolutions.

2 Newcomb's Problem

Robert Nozick describes Newcomb's problem as follows:

> A being in whose power to predict your choices correctly you have great confidence is going to predict your choice in the following situation. There are two boxes, B1 and B2. Box B1 [which we can assume is transparent] contains $1,000; box B2 [which is opaque] contains either $1,000,000 ($M) or nothing. You have a choice between two actions: (1) taking what is in both boxes; (2) taking only what is in the second box. Furthermore, you know, and the being knows you know, and so on, that if the being predicts you will take what is in both boxes, he does not put the $M in the second box; if the being predicts you will take only what is in the second box he does put the $M in the second box. First the being makes his prediction; then he puts the $M in the second box or not, according to his prediction; then you make your choice. (1993: 41)

Given that, once it's time for you to make your choice, the amount of money in each box has already been set, and given that you prefer more money to less, if rationality prohibits you from showing standard-form restraint, and so prohibits you from passing up a benefit while knowing that doing so will bring about an outcome that you disprefer to another outcome that you could bring about instead, you will, if you are rational, take both boxes and not leave the $1000 in the first box behind. But, insofar as your mindset disposes you to avoid showing standard-form restraint, and so to take both boxes, the predictor will presumably predict that you will take both boxes and so will not put $M in the second box. If, by contrast, your mindset disposes you to show standard-form restraint and pass up the first box, even though you prefer gaining the fixed amount of money that is now in the second box plus the $1000 in the first box over gaining just the fixed amount that is now in the second box, the predictor will presumably predict that you will take only the second box and fill it with $M. The person who is disposed to show standard-form restraint in Newcomb-type cases can thus expect to do better than the person who is disposed not to show such restraint (assuming there is no confusion about the situation). Shall we conclude that rationality is a curse in such cases, or shall we instead allow that, in such cases, the (instrumentally) rational agent is disposed to show standard-form restraint?

3 Practical Rationality and Practical Success

If one thinks of practical rationality as a mindset that suits one for practical success, then it seems like we should allow that the rational agent is some-times disposed to show standard-form restraint. Why preclude a beneficial disposition from the mindset of the rational agent? The "one-boxer" who is accused of irrationality by the "two-boxer" can, it seems, forcefully respond with the stock retort "If you're so rational, why aren't you rich?"

But, intuitively, there seem to be cases, or at least possible cases, in which one's mindset, though it suits one for practical success, also makes one at least somewhat irrational. Suppose, for example, that one is generally bene-fited by self-enhancement motivations that skew one's beliefs about one's likeability, competence, and attractiveness. Although this mindset makes one better suited to success than agents with a more objective mindset, it also seems to make one *less* rational. Here, it might be responded that while the mindset makes one less *theoretically* rational, it does not make one less *practically* rational. Expanding on this suggestion, it might be boldly claimed that, although things like inconsistency, myopic planning, and distorted thinking are condemned by theoretical rationality, they can be perfectly suitable when it comes to practical rationality. This claim is quite radical, and the one-boxer can plausibly insist that she need not defend it. For she can maintain that, while some dispositions involve inconsistency, myopic planning, and distortions, and are thus practically irrational, the disposition to show standard-form restraint in Newcomb-type cases need not involve any such failures.

Certainly some one-boxers might be disposed to choose one box in Newcomb-type cases due to distortion. Perhaps I'm disposed to choose one box because I just can't grasp that my choosing one box will not retroactively get $M into the box. (In this case, my choosing one box does not involve my showing standard-form restraint, since I believe that forfeiting the second box will bring about very handsome compensation indeed.) But a one-boxer need not have a distorted view of her situation. She may just have a different way of reasoning from her preferences and the facts. She might, for example, seek to maximize "evidentially expected utility" rather than causally expected utility (Nozick 1993: 43). Otherwise put, she might seek to perform the act that, roughly speaking, is best in terms of what outcome it is evidence for, rather than the act that is best in terms of what outcome it causes. If so, then, if she understands her situation correctly, she will pick one box rather than two. For, although picking two boxes is best in terms of what outcome it causes, since picking two boxes gets her $1000 more than picking one box, picking one box is best in terms of what outcome it is evidence for, since picking one box is evidence that she will gain $M, while picking two boxes is evidence that she will gain a mere $1000. This one-boxer need not be inconsistent, myopic, or delusional. And she does show standard-form restraint, since she passes up the first box even though she prefers gaining the fixed amount of money that is now in the second box plus the $1000 in the first box over gaining just the fixed amount that is now in the second box. And, of course, as a one-boxer, she is better suited for success in Newcomb-type cases than an agent who is disposed not to show standard-form restraint and to, instead, take both boxes. (Notably, and as should by now be clear, exhibiting standard-form restraint is consistent with exhibiting no restraint at all with respect to maximizing evidentially expected utility. Relatedly, since maximizing both causally expected utility and evidentially expected utility is not always possible, every choice procedure will call for showing restraint with respect to at least one of these standards in at least some cases, and so the mere fact that a procedure calls for a certain form of restraint is certainly not a sufficient basis for dismissing the procedure as irrational. If standard-form restraint is irrational, it is not just because it qualifies as a form of restraint.)

4 The Toxin Puzzle

The idea that a willingness to show standard-form restraint in certain circumstances might be rational because being so disposed can serve one's preferences better than a blanket unwillingness to show standard-form restraint is familiar not just from responses to Newcomb's problem, but also from responses to Gregory Kavka's famous "toxin puzzle" (whose relation to Newcomb's problem has not gone unnoticed; see, e.g., Andreou 2008; Greene 2013: §2.3). In Kavka's invented case, an eccentric billionaire... places before you a vial of toxin... [You are provided with the following information:] If you drink [the toxin], [it] will make you painfully ill for a day, but will not threaten your life or have any lasting effects... The billionaire will pay you one million dollars tomorrow morning if, at midnight tonight, you intend to drink the toxin tomorrow afternoon... You need not drink the toxin to receive the money; in fact, the money will already be in your bank account hours before the time for drinking it arrives, if you succeed... [The] arrangement of... external incentives is ruled out, as are such alternative gimmicks as hiring a hypno-tist to implant the intention... (1983: 33–4)

If we accept the standard view that the (instrumentally) rational agent does not show standard-form restraint, we again seem to arrive at the conclusion that rationality is a curse. For, given the standard view, the rational agent will never incur a cost while knowing that doing so will bring about an outcome that she disprefers to another outcome that she could bring about instead. It follows, assuming that rational agents are not myopic, that a rational agent will be confident that he will not drink the toxin. But then a rational agent will not be able to form the intention to drink the toxin. He will thus not be able to get the million. An agent capable of showing standard-form restraint need not have this problem. For, insofar as an agent is capable of showing such restraint, his recognition that drinking the toxin will require showing such restraint need not generate confidence that he will not drink the toxin.

As in the case of Newcomb's problem, getting rich need not involve inconsistency, myopia, or distorted thinking. Although I might get the million because, for example, I can't grasp that I don't have to drink the toxin to get the million, I might instead get it because I have a different way of reasoning from the facts and my preferences. Might a disposition to maximize evidentially expected utility rather than causally expected utility again do the trick? Not in this case. To see this, suppose I have the dispos-ition to maximize evidentially expected utility and I know this. I will then know that I will not drink the toxin when the time arrives. For when that time arrives, I will already know whether I got the million dollars or not, and so, at that point, neither my drinking the toxin nor my refraining from drinking the toxin will be evidentially relevant with respect to the question of the million. The expected evidential utility of my drinking the toxin will be one day of sickness; and the expected evidential utility of my not drinking the toxin will be the status quo. Since adding one day of sickness worsens the status quo, I know in advance that I will not drink the toxin. I will thus not be able to form the intention to drink the toxin, and so I will not be able to gain the million.

A somewhat different disposition will, however, do the trick. This is the disposition of "constrained maximization." Like the maximizer of causally expected utility, the constrained maximizer is concerned with causal impact. But unlike the maximizer of causally expected utility, the constrained maximizer is, roughly speaking, disposed to constrain her maximizing to the set of actions that accord with her prior intentions (Gauthier 1994). Putting aside a qualification that I will get to below (which allows for some instances of legitimately abandoning prior intentions), the basic idea is that if W, X, Y, Z are the actions available to me, and if W and X accord with my prior intentions, but Y and Z do not, then I will limit my maximizing to the set W and X, putting aside as ineligible Y and Z because they do not accord with my prior intentions.

Suppose a constrained maximizer finds herself in a toxin case. Can she, assuming she correctly understands her situation, get the million? The answer seems to be yes. For, the constrained maximizer can be confident that, if she forms the intention to drink the toxin, she will drink it, since not drinking it would not accord with her prior intention. There is thus no obstacle to her forming the intention to drink the toxin and getting the million (Gauthier 1994).

One problem that has been noted with the basic, unqualified form of constrained maximization described above is that constrained maximizers that follow this basic form of constrained maximization (henceforth "basic constrained maximizers") can suffer from being "too resolute" (Gauthier 1994). Suppose that I am a basic constrained maximizer and find myself in what appears to be a toxin case. I thus form the intention to drink the toxin. The next day, I find out the whole thing was some kind of scam. There is no million in my bank account. Indeed, some of my funds have been removed. As a basic constrained maximizer, I will still go ahead and drink the toxin, which seems clearly irrational. For I am here constraining my behavior to accord with an intention that I did not even benefit from forming. The basic constrained maximizer's disposition needs refinement, so that (roughly put) one is not bound by prior intentions that falsely appeared beneficial due to misrepresentation of the situation. With some refinement along these lines, one's disposition will not leave one vulnerable in scam toxin cases, but will still allow one to get the million in genuine toxin cases (in which one *does* benefit from forming the intention to drink the toxin).

5 Constrained Maximization, Direct Evaluation, and Indirect Evaluation

Given its focus on causal, rather than evidential impact, the idea that rationality requires (refined) constrained maximization may seem less of a departure from the standard view than the idea that rationality requires maximizing evidentially expected utility. And, indeed, according to one way of casting the difference between the standard approach and the revisionary approach favoring constrained maximization (henceforth "the revisionary approach" for short), the difference is not in the form of maximization (which is causal in both cases), but in its primary object: the standard approach asks what action best serves the agent's concerns, whereas the revisionary approach asks what deliberative disposition best serves the agent's concerns. The standard approach assumes that the direct object of evaluation should be actions and that the best deliberative disposition is just the one that selects maximizing actions, whereas the revisionary approach suggests that the direct object of evaluation should be deliberative dispositions and that the best action is just the one that is selected by the maximizing deliberative disposition. (Figuring as common ground is the assumption that the standards of rational advisability can be accepted without inconsistency, and so cannot include standards for actions and for deliberative dispositions or procedures that are not coordinated with one another, wherein, for example, one standard endorses a deliberative disposition or procedure that selects action X as all-things-considered required while another standard directly dismisses the selection of action X as all-things-considered impermissible.) The revisionary approach defends its suggestion that the direct object of evaluation should be deliberative dispositions via the idea that, in relation to serving one's concerns well, the key is not maximizing at the level of actions, but maximizing at the level of deliberative dispositions (with adherence to the maximizing deliberative disposition sometimes prohibiting the selection of the maximizing action). The standard assumption that the direct object of evaluation should be actions is critiqued as nothing more than an undefended assumption – one that seems safe and uncontentious until one thinks about cases like the toxin case (Gauthier 1994).

6 In the Neighborhood of the Newcomb-predictor

For those who are warming up to the idea that constrained maximization might be rational, but are still skeptical about the rationality of maximizing evidentially expected utility, a variation on Newcomb's problem will be of interest. In this variation, one can defend the one-boxer without favoring an evidential approach over a causal approach. One need only recognize constrained maximization as the (causally) maximizing deliberative disposition, and accept the revisionary suggestion that actions should be selected on the basis of whether they are supported by the maximizing deliberative procedure. The relevant variation is David Gauthier's version, according to which you know in advance that you are "in the neighbourhood" of a "Newcomb-predictor." Given this knowledge (or even just the anticipation that you might at some point be in the neighborhood of a Newcomb-predictor), you can form the intention to take only one box if the Newcomb-predictor ever decides to make you his trademark offer. If you later find yourself in the relevant situation, the Newcomb-predictor, aware of your resoluteness, will put a million in the box, and you, as a constrained maximizer, will stick with your beneficial intention to take one box.

Of course, a constrained maximizer in the original Newcomb case, in which the agent does not anticipate the possibility of being approached by a Newcomb-predictor, will not be able to gain the million. For, not anticipating the offer, she will not form any intention in advance. And, not having formed any intention in advance, her later choice will not be constrained by a prior beneficial intention to select only one box. She will thus maximize over the two options and thus take both boxes. Since the Newcomb-predictor will have anticipated this, there will be no million in the second box.

7 The Newxin Puzzle

Another interesting variation on Newcomb's problem leaves both the constrained maximizer and the evidential expected utility maximizer at a loss (Gibbard and Harper 1981). The variation appears, in my work, as a hybrid of Newcomb's problem and the toxin case, and so I dub it the "Newxin puzzle" (2008). In the variation in question, the procedure is as follows:

Before you have any idea what is going on (and before you ever consider the possibility that you might be in the neighborhood of a Newcomb-predictor), the being makes his prediction. He puts $M in the opaque box if he predicts that you will take only the opaque box; he puts $0 in the opaque box if he predicts that you will take both boxes. He then lets you know what is going on, without telling you his prediction. You then open the opaque box, and see whether it contains $M or not. Finally you get to take the $1000 box or not.

While a certain type of refined resolute agent that I will get to shortly can form the intention to pass up the $1000 transparent box after opening the first box, and then follow through, the constrained maximizer that we have been considering cannot. For, given that, as in the original Newcomb case, when the agent has the option of forming her intention, the predictor has already made his prediction and either put the $M in the opaque box or not, forming the intention to pass up the transparent box does not bring about (or even increase the likelihood) of any benefit. It is thus clear that the constrained maximizer would not follow through on any such intention (Andreou 2008). Indeed, from the point of view of the constrained maximizer, the Newxin case is not significantly different from the two-transparent-boxes version of New-comb's problem (discussed in, for example, Meacham 2010), wherein the agent is presented with the boxes at the same time (as in the original New-comb case) but (in this variation) *both* boxes are transparent. (Note that, as in the original Newcomb case and the Newxin case, I assume that the agent is presented with the boxes before ever considering the possibility that he might be in the neighborhood of a Newcomb-predictor.)

The maximizer of evidentially expected utility will also end up without the $M in both the Newxin case and in the two-transparent-boxes version of Newcomb's problem. For, when the maximizer of evidentially expected utility is faced with the possibility of showing standard-form restraint and passing up the $1000 box, evidentially expected utility will not speak in favor of doing so. Already knowing whether or not there is $M in the other box, the choice regarding whether to also take the $1000 box or not is evidentially irrelevant with respect to the question of the million. Choosing to take the $1000 box is simply evidence that I will end up with an extra $1000. Choosing not to take it is simply evidence that I will not get the extra $1000. The evidentially expected utility maximizer will thus favor taking both boxes and so will presumably not be provided with a million-dollar box to begin with.

Notably, the Newxin case and the two-transparent-boxes version of New-comb's problem are importantly different from the point of view of a certain type of refined resolute agent. Consider an agent who is resolute with respect to an intention that calls for standard-form restraint if and only if the formation of the intention maximizes causally expected utility *or* evidentially expected utility. A resolute agent of this sort will not be able to gain the million in the two-transparent-boxes version of Newcomb's problem, since, when both boxes are transparent, forming the intention to pass up the $1000 box calls for standard-form restraint but maximizes neither causally expected utility nor evidentially expected utility. She will, however, be able to gain the million in the Newxin case, since forming the intention to pass up the $1000 box does, in that case, maximize evidentially expected utility, and so the intention can be formed and followed through on, given this agent's brand of resoluteness.

Where do we go from here? Perhaps (and this suggestion is admittedly highly speculative) Newcomb's problem, the Newxin case, and the two-transparent-boxes version of Newcomb's problem can all be addressed via an idea that fits with the spirit of constrained maximization, but extends to beneficial predictions (where the question of whether a prediction is beneficial differs from the question of whether the predicted occurrence is beneficial, just as the question of whether an intention is beneficial differs from the question of whether the intended action is beneficial). Suppose that I am disposed to constrain my behavior not only in accordance with beneficial intentions, but also in accordance with beneficial predictions. Such a deliberative disposition would be very beneficial in Newcomb-type cases. In particular, in all three Newcomb-type cases under consideration, when the opportunity to take both boxes arrives, I will reason as follows: I benefit from the prediction that I will take only one box, so I will take only one box. Since the predictor can anticipate that I will reason in this way, he will put the million in the appropriate box and I will get it.

Like the original, basic version of constrained maximization, this disposition needs refinement. Notice, for example, that if one is dealing not with an extremely talented predictor, but a thoroughly incompetent predictor, then one does not benefit from having the disposition to constrain one's behavior in accordance with beneficial predictions. Suppose, for example, that the predictor is so lousy at making predictions that he decides to just base his predictions on a coin flip. Heads, he predicts you will take one box and puts the million in; tails, he predicts you will take two boxes and doesn't put the million in. In this case, someone who has the disposition to constrain his behavior in accordance with beneficial predictions does worse than someone who does not. For, having the disposition does not help get the million in the box, and not having it gains one an extra $1000 dollars.

What about the disposition to constrain one's behavior in accordance with beneficial predictions insofar as one is (appropriately) confident one is dealing with a predictor who has the power to correctly predict one's choices? While this refinement seems to be on the right track, things are complicated. Suppose I have this disposition and find myself in the Newxin case. Suppose further that I have opened the first box and it contained $M. I therefore know that the predictor predicted that I would not take the second box. How will

I reason when I have the chance to take the second box? I can certainly say that the prediction that I will take one box is beneficial. I will then have to consider whether I am confident that I am dealing with a predictor who has the power to correctly predict my choices. By hypothesis, I start out with this confidence. But can I maintain my confidence in the being's power to correctly predict my choices while knowing that, if I take the second box, I will refute the being's prediction? The answer, it seems, is "so long as I know, or at least have good reason to believe, that I will not take the second box." But, given my deliberative disposition, can I know, or at least have good reason to believe, that I will not take the second box? The answer, it seems, is "so long as I can be confident the being has the power to correctly predict my choices." But can I maintain my confidence in the being's power to correctly predict my choices while knowing that, if I take the second box, I will refute the being's prediction? I have, it seems, come full circle and still have no answer to the question raised by my deliberative disposition.

8 Rationality, Restraint, and the Puzzle of the Self-torturer

We have been exploring the possibility that rationality might involve standard-form restraint. But consider the following objection by a defender of the view that one should never incur a cost or pass up a benefit if one knows that doing so will bring about an outcome that one disprefers to another outcome that one could bring about instead: Both Newcomb's problem and the toxin puzzle are contrived cases in which the agent is being rewarded for being irrational. They do not provide us with any reason to question the rationality of refusing to show standard-form restraint. For there is nothing odd about the conclusion that, when things are contrived so that one is rewarded for being irrational, rationality is a curse. To really support the rationality of standard-form restraint, one needs a case in which such restraint seems called for, but not because the case has been contrived so that the agent is directly rewarded for being prepared to show such restraint.

This brings us to another interesting case in the debate concerning rationality and restraint, namely Warren Quinn's case of the self-torturer. Quinn describes the case as follows:

> Suppose there is a medical device that enables doctors to apply electric current to the body in increments so tiny that the patient cannot feel them. The device has 1001 settings: 0 (off) and 1... 1000. Suppose someone (call him the self-torturer) agrees to have the device, in some conveniently

portable form, attached to him in return for the following conditions: The device is initially set at 0. At the start of each week he is allowed a period of free experimentation in which he may try out and compare different settings, after which the dial is returned to its previous position. At any other time, he has only two options—to stay put or to advance the dial one setting. But he may advance only one step each week, and he may *never* retreat. *At each advance he gets $10,000.*

Since the self-torturer cannot feel any difference in comfort between adjacent settings [or at least he cannot, with any confidence, determine whether he has moved up a setting just by the way he feels], he appears to have a clear and repeatable reason to increase the voltage each week. The trouble is that there *are* noticeable differences in comfort between settings that are sufficiently far apart. Indeed, if he keeps advancing, he can see that he will eventually reach settings that will be so painful that he would then gladly relinquish his fortune and return to 0. (1990: 198)

(Note that, for reasons I won't delve into here, I will assume that the option of dialing up one setting each week will expire after 1000 weeks, and that in earlier weeks, the self-torturer's declining to dial up a setting that week does not remove the option of dialing up a setting the following week.) Quinn assumes that, for every n between 0 and 999, the self-torturer, quite reasonably, prefers his situation at setting $n+1$ to his situation at setting n. If this is an acceptable hypothesis, then we get the following "reductio": Suppose that rationality prohibits an agent from passing up a benefit if doing so will bring about an outcome that he disprefers to another outcome that he could bring about instead. And let n be some setting between 0 and 999. Then rationality prohibits the self-torturer from remaining at setting n given the option of proceeding to setting $n+1$. For, however many *other* dial up opportunities he takes, taking this one gets him an extra $10,000 without any noticeable difference in pain. But if the self-torturer is required, for every n between 0 and 999, to proceed to setting $n+1$, then he will end up at 1000 and never be able to retreat. And, it seems intuitively clear that it would be irrational for the self-torturer to proceed all the way to 1000. Rationality must, it seems, require that the self-torturer show standard-form restraint and, at some point before he reaches setting 1000, voluntarily pass up a benefit even though he knows that doing so will bring about an outcome that he disprefers to another outcome that he could bring about instead.

While this case is contrived in certain ways, it is not one in which the agent is directly rewarded for being prepared to show standard-form restraint.

Although the self-torturer can end up significantly better off if he is prepared to show standard-form restraint than if he is not, this is not because someone is gauging whether the self-torturer is prepared to show standard-form restraint and rewarding him if he is. And indeed, there are realistic cases that are structurally similar to the case of the self-torturer in which no interested third party need be involved at all. As Quinn emphasizes, "the self-torturer is not alone in his predicament" (1993: 199) – the case resembles many cases of temptation, including some in which no one else plays an important role. It may be, for example, that adding one more "pleasurable moment of idleness" will not take me from a productive life to a "wasted" life, but adding many such moments will (ibid.: 199). Relatedly, I may, at each choice point, prefer adding one more pleasurable moment of idleness, but also prefer a productive life to a wasted one. Here, the need to show standard-form restraint seems both clear and familiar.

9 Conclusion

Newcomb's problem can be located among a set of intriguing philosophical cases that have generated a great deal of interesting debate concerning rationality and restraint. It might seem like, at least when it comes to instrumental rationality, there is no room for standard-form restraint. If what matters is serving one's preferences well, how can it make sense to knowingly incur a cost or pass up a benefit while knowing that doing so will bring about an outcome that one disprefers to another outcome that one could bring about instead? As we have seen, in light of cases like Newcomb's problem, the toxin case, and the puzzle of the self-torturer, this seemingly rhetorical question has been taken up in earnest. Sophisticated philosophical arguments have been developed based on the idea that there is a close connection between practical rationality and practical success. Consensus has not been reached, but the debate has certainly led to a variety of nuanced and insightful contributions.

3   The "Why Ain'cha Rich?" Argument

Arif Ahmed

In Newcomb's Problem you must choose between two acts, A_1 ("one-boxing") and A_2 ("two-boxing"); the state of the world is either S_1 ("predicted one-boxing") or S_2 ("predicted two-boxing"); and your payoffs, which we can assume to be in dollars, are as follows:

Table 1: Newcomb's Problem

	S_1	S_2
A_1	M	0
A_2	$M + K$	K

(In this table $M = 1$ million and $K = 1000$.) Your choice between A_1 and A_2 has no *effect* on which of S_1 and S_2 is true. But your choice is *symptomatic* of the state: we know that in the past, and we have every reason to expect that now and in the future, choosing A_1 is strongly correlated with S_1, and choosing A_2 is strongly correlated with S_2. That is the basic structure of Newcomb's Problem. Suppose that you are facing such a problem. What should you do?

Here is a flat-footed answer. Almost every time anybody chooses A_1 she gets M. Almost every time anybody chooses A_2 she gets K. You greatly prefer M to K. So, choose A_1!

That essentially statistical argument, known in the literature as "Why Ain'cha Rich?" (WAR), has clear intuitive force. It is hard to contest the premises. The precise figures may vary in different realizations of the Newcomb scenario, but it is safe to say this: we know that the vast majority of those who choose A_1 get M, and that a similar proportion of those who choose A_2 get K. In the limiting case we can even assume that *everyone* (including you) who chooses A_1 gets M, and that everyone who chooses A_2 gets K. And it is given that you greatly prefer M to K.

As for whether the premises support the conclusion: *prima facie*, they do. We make practical inferences like this all the time, as do many animals at least to some extent;[1] had this not been so, nobody would ever have learnt anything. If anything is ever a *prima facie* reason for anything else, then "Whenever I put my hand near the fire I get burnt" is a *prima facie* reason not to put your hand near the fire, a reason that invokes the same statistical type of reasoning as does WAR.

But the inference is not logically valid, and so, given additional assumptions, it may not even be reasonable. "This food is nutritious, so I'll have some" looks reasonable; but it stops looking that way when you realize that this food is also very expensive. The question is therefore whether there are additional facts about Newcomb's Problem that should stop us from finding WAR persuasive if we keep them in mind.

In this chapter I'll consider and reject several such putative facts. They are (1) that the statistical advantage of one-boxing vanishes when we restrict attention to cases matching known features of your situation; (2) that it does so when we restrict attention to cases matching unknown features of your situation; (3) that it does so when we restrict attention to cases matching unknown features of your situation that are causally independent of what you do; (4) that it does so when we restrict attention to cases in which the agent shares your opportunities for gain.[2]

1 Restriction on a Known Feature

Apples are generally nutritious; suppose 81% of them are. Boiled sweets are generally not nutritious; suppose 9% of them are. Here is an apple and here is a boiled sweet. You greatly prefer eating nutritious food. So, eat the apple!

---

[1] For instance, Balleine *et al.* "initially fed rats on an orange–lime drink that was sweetened with sugar before giving one group the opportunity to sample an unsweetened lime-alone drink. When subsequently tested with an unsweetened orange solution, they drank more than control rats for which the lime-alone drink had been sweet" (Dickinson 2012, citing Balleine *et al.* 2005). On the other hand, rats "even after thousands of trials fail to learn the link between a stimulus and a reward, and there is no secondary reinforcer... Humans appear to be less constrained and can (though may not) link virtually anything to anything else across time" (Suddendorf and Corballis 2007: 311).

[2] A different sort of objection, that I lack the space to consider here but hope to discuss at greater length elsewhere, trades on the distinction between rational *acts* and beneficial *dispositions*. According to this line, the difference in wealth (*ceteris paribus*) between those who one-box and those who two-box does not show that the former act is more rational, but only that the predictor pre-rewards those with the disposition to perform it. (There are traces of this idea in Lewis 1981b.) I am willing to grant that the predictor rewards those with the disposition to one-box, but here I focus only on whether the statistics underlying this fact *also* make one-boxing the rational act. (Thanks to a referee.)

One thing that would undercut that practical reasoning is the additional premise that *this* apple is of the experimentally bred X-type, of which only 1% are nutritious. Knowing that would undercut the original argument; it would indeed make it rational to eat the boiled sweet rather than the apple, if you are going to eat either. What does the undercutting is the placing of the apple into a strictly more exclusive class ("X-type apple") than that of "apple," within which the incidence of nutritious items is very much sparser.

Similarly, an insurance company once wrote to the philosopher Nancy L. D. Cartwright in the following terms:

> It simply wouldn't be true to say, "Nancy L. D. Cartwright… if you own a TIAA life insurance policy you'll live longer." But it is a fact, nonetheless, that persons insured by TIAA do enjoy longer lifetimes, on the average, than persons insured by commercial insurance companies that serve the general public.[3]

That is hardly a good reason for Prof. Cartwright or anyone else to purchase life insurance from TIAA rather than from "commercial insurance companies that serve the general public." The reason is that it is very easy to determine facts about one's situation that place one in a class within which this correlation breaks down. For instance, if one has a rough idea of one's general health, one's dietary and exercising habits, and one's pattern of alcohol and tobacco consumption, it is very likely that one knows enough to make the correlation irrelevant. It may be true for various reasons that longevity amongst people in general is correlated with being a customer of TIAA.[4] But it is almost certainly *not* true that longevity amongst people with your health, diet, etc., is correlated with being a customer of TIAA.

Can we undercut WAR in the same way? That is, can we find some smaller class C of events that knowably encompasses one of the options in Newcomb's Problem, such that the average return on C-type acts is very different from that envisaged in WAR?

One tempting line of thought arises from contrasts like this: (a) 99% of the time when my clock strikes midnight, so does yours. But (b) when I *bring it about* (for instance, by directly adjusting the mechanism) that my clock strikes midnight, yours does *not*. This contrast witnesses the general principle

---

[3] Cartwright 1979: 420.

[4] For instance, this may be because TIAA has a high credit rating, and people who are careful enough with their money to research credit ratings are also careful enough with their health not to take unnecessary risks.

that for many natural values of P and X, the incidence of a property P within a class of events X differs greatly from the incidence of P within the class $X^*(\subset X)$ of events in that class that someone has directly brought about. Reflection on principles like this and examples like that gives rise to two ideas. First, if we look only at the class A_1^* of A_1-type events that somebody has brought about directly, we might find that within *this* class the average return for A_1 is significantly lower than that envisaged by WAR. And if we look only at the class A_2^* of A_2-type events that somebody has brought about directly, we might find that within *this* class the average return for A_2 is significantly *higher* than that envisaged by WAR. Second, A_1^* and A_2^* certainly are classes to which your options A_1 and A_2 would respectively belong, if you were to realize them. So perhaps the statistical advantage of A_1 over A_2 would be lost if we focused on this subclass, and in a way that undercuts WAR.

That line of thought gets us nowhere. It is *definitive* of Newcomb's Problem that the accuracy of the prediction is high even *within* the classes of cases in which A_1 and A_2 are actions, that is, events that somebody (for instance, you) has directly brought about. Even within this class, the return for A_1 is close to M and that for A_2 is close to K.

In any case, suppose that we find two classes A_1^* and A_2^* within which your options fall, such that the reliability of the predictor within at least one class is relatively low. For instance, suppose that you know that your current mental state (your beliefs, your desires, your current preference for one option over another) satisfies a condition R. Let R_1 be the class of A_1-type choices by individuals whose mental states satisfy R. Let R_2 be the class of A_2-type choices that satisfy R. And suppose that the prediction is based on whether your mental state satisfies R. Then the predictions of actions within R_1 or R_2 will be highly *in*accurate, since the prediction will be the same (or will have the same distribution) in both classes. This *will* affect the average returns: specifically, the average return for A_2 will exceed that for A_1 by K. But then we no longer have a Newcomb case at all. For since you know that your current mental state satisfies R, there is no basis for regarding what you do as symptomatic of what you were predicted to do. This line of argument might work in defense of Evidential Decision Theory – and it has indeed been proposed in this connection under the label "Tickle Defense" – but it has no tendency to diminish the force of WAR as applied to Newcomb's Problem itself.[5]

---

[5] For the "Tickle Defense," see, e.g., Horgan 1981: 178–9; Eells 1982 ch. 6–7; the chapter in this volume by Bermúdez. For the point that these considerations undermine the Newcomb-like character of the problem, see Nozick 1969: 219 n. 11.

2 Restriction on an Unknown Feature

In any large totality of actual trials of the Newcomb Problem, we expect that A_1 does better on average than A_2. But we can segregate the trials into two exhaustive and exclusive subclasses, in *each* of which A_2 does better than A_1. Consider the subclass S_1 of cases where the agent is predicted to choose A_1. Within S_1, agents who choose A_1 always get M and agents who choose A_2 always get $M + K$: so there the return for A_2 exceeds the return for A_1 by K. Consider the subclass S_2 of cases where the agent is predicted to choose A_2. Within S_2, agents who choose A_1 get nothing and those who choose A_2 get K: so there too the return for A_2 exceeds the return for A_1 by K. S_1 and S_2 are exclusive classes that jointly exhaust the class of all trials.

The statistics governing Newcomb's Problem therefore make it an example of the more general statistical phenomenon known as Simpson's Paradox. A body of statistics is Simpson-paradoxical if a correlation that obtains within a population is reversed within each cell of some segregation (or "partition") of that population. One famous real-life example concerns the relative effectiveness of two types of treatment for kidney stones.[6] Overall treatment A (minimally invasive) was statistically more effective than treatment B (open surgery), that is, the choice of treatment A was positively correlated with recovery. But restriction to either of the two types of kidney stone (small or large) reverses this correlation: treatment B was more effective than treatment A for the treatment of small kidney stones, and treatment B was more effective than treatment A for the treatment of large kidney stones. The obvious explanation of this is that open surgery was preferred for the more difficult-to-treat large kidney stones.

This fact may seem to support a statistical argument for two-boxing. It's true that when you are choosing, you don't know what the prediction is. But you do know (a) that two-boxing does better than one-boxing when one-boxing was predicted; (b) that two-boxing does better than one-boxing when two-boxing was predicted. And whether or not there is a statistical argument on these grounds *in favor of* two-boxing, one might well think that those grounds themselves undercut the WAR argument for one-boxing.

The difficulty with that line is that it proves too much. Suppose that you are ill and wondering whether to take some medicine. The medicine costs $1. You know that almost everyone in your condition who takes the medicine eventually recovers; almost nobody who does not take the medicine eventually

[6] Charig *et al.* 1986.

recovers. And you have every reason to expect this pattern to continue. This seems a good argument for taking the medicine. But the foregoing objection to WAR would apply here too, since we can segregate the class of trials into two exhaustive and exclusive subclasses, in *each* of which eschewing the medicine does better than taking it. Consider the class S_1 of cases where the agent recovers. Within S_1, agents who take the medicine always recover but always lose \$1, and agents who eschew the medicine always recover without paying \$1. Within S_1 then, the return for eschewing the medicine exceeds the return for taking it by \$1. Consider next the class S_2 of cases where the agent does *not* recover. Within S_2, anyone who takes the medicine loses \$1 but remains ill, and anyone who does not take the medicine also remains ill but keeps the \$1. Here too, the return for eschewing the medicine exceeds the return for taking it by \$1. And S_1 and S_2 are exclusive classes that jointly exhaust all trials.

If this kind of reasoning undercut our grounds for one-boxing, then it undercuts our grounds for taking the medicine. But not taking the medicine for this reason would be absurd, and applied consistently, this sort of fatalism rules out incurring any cost at all in the pursuit of any good at all, for we could always segregate the statistics into classes within each of which the correlation between the incurring of the cost and the achievement of higher utility was reversed. So far then, we can say this: the mere existence of a segregation of Newcomb cases into classes, in each of which two-boxing does better than one-boxing, is *not* by itself grounds for rejecting WAR.

3 Restriction on a Causally Independent Feature

The immediate response is that there is a relevant difference between the fatalist reasoning and the reasoning that sought to undercut WAR. In the latter case, but not the former, the segregation of trials is into classes with the following property: which class you are in is *causally independent* of what you choose. Whether you take the medicine *does* have a causal impact on whether your case falls into the class of cases in which the agent recovers or the classes of cases in which the agent does not. But in the Newcomb case, your choice *does not* thus affect whether your case falls among those trials in which the agent was predicted to one-box or among those in which the agent was predicted to two-box.

This illustrates a popular attitude towards cases involving Simpson's Paradox more generally. Put very simply, it is this: suppose that we have a segregation of trials into classes in which the correlation between action and success is reversed. Then this is practically relevant if and only if the choice of action is causally *irrelevant* to which element of the *segregation* is actual. We find examples of this attitude amongst computer scientists,[7] philosophers[8] and statisticians.[9]

This way of thinking about the practical implications of statistical evidence is certainly intuitive for people who – like most of us – find it entirely natural to talk about causality. But setting intuition aside, it is not clear *why* it should move anyone. Go back to the imaginary situation of someone who is capable of inductive but has not yet learnt about causal reasoning. Like an infant or an animal, he extrapolates from observed or otherwise known patterns of co-occurrence and succession, and on this basis chooses *A* over *B* whenever *A* has a projectibly better track record in situations that match known features of the present situation. This doubtless naïve agent will prefer one-boxing. Now we somehow introduce him to our *causal* vocabulary and point out that across situations that match the actual situation in some *unknown* respect that is *causally* independent of his choice, two-boxing does better than one-boxing. The natural response is then: so what?

Suppose someone insisted on applying a *health-based* criterion of rational choice to any correlation-reversing segregation of relevant trials. This is as follows: if, within each relevant class, all trials agree on the agent's state of health at some specified time or within some specified interval of time, then the partition is practically relevant; otherwise it is not. This health-based criterion would recommend not taking the medicine in the example in section 2. For we can segregate the set of trials into two classes such that all agents recover from the illness in one class and no agents recover in the other. And within each class the agents that take the medicine are worse off (by $1) than those that do not.

In response to that we can and should ask: what is so special about this kind of segregation that gives it any practical relevance? What is it about the state of the world *regarding the agent's health*, in virtue of which an option *A* is to be preferred to an option *B* whenever *A* does better than *B* if we control for health?

And we can and should pass from that defensive move to an offensive one: the health-based criterion is silly, at least as applied to the medical example, because we know that on average agents who rely on it in these cases end up worse off, both overall and as regards their health in particular, than agents

---

[7] Pearl 2014.   [8] Meek and Glymour 1994.   [9] Lindley and Novick 1981.

who instead rely on the overall correlation between taking the medicine and recovering from the condition.

Equally then, when the advocate of the causal independence criterion presents it as an argument for two-boxing, we can and should ask just what is so special about the partitions over trials that she regards as having practical relevance? What is so special about the state of the world *regarding factors that are causally independent of the agent's choice*, in virtue of which an option A is to be preferred to an option B whenever A does better than B if we control for those factors? And we can and should pass from that defensive move to an offensive one: the causal independence criterion is silly, at least as applied to Newcomb's Problem, because we know that on average agents who rely on it in these cases (and so take both boxes), will end up worse off, both overall and as regards the prediction, than agents who instead rely on the overall correlation between one-boxing and getting rich.

Is anything special about the relation of causal independence that makes practically relevant the fact that A_2 always performs better than A_1 across trials that hold fixed causally independent factors? The answer presumably depends on an account of what causal independence is supposed to be. There are as many answers to that question as there are analyses of causality itself. The trouble is that the same question arises with regard to these analyses: given an analysis equating causal independence with some independently comprehensible relation R, we can always ask for, but never seem to get an answer to, the question: what is it about R that confers special practical relevance on the fact that A_2 always performs better than A_1 across trials that hold fixed all factors that are R-related to the choice between them?[10]

---

[10] Some theories of causality are exceptions to this general point. Consider the agency theory of Menzies and Price (1993), in which to say that A causally promotes B is to express the positive subjective probabilistic relevance of one's own direct bringing-about of A to the occurrence of B. More formally, if Cr is the probability function that represents my beliefs, and if A^* is the proposition that I directly bring it about that A, then what constitutes my belief that A causally promotes B is that $Cr(B|A^*) > Cr(B)$. On this theory, however, the partition $\{S_1, S_2\}$ of states in Newcomb's Problem is *not* causally independent of my choice between A_1 and A_2, since the definition of Newcomb's Problem implies that $Cr(S_i|A_i^*) = Cr(S_i|A_i) > Cr(S_i)$ if $i = 1,2$. So on this theory, the purported argument from causal independence does not get off the ground, and in fact Causal Decision Theory in conjunction with the Menzies-Price theory appears to recommend A_1. Again, consider the interventionist theory of Woodward (2003), according to which (roughly) a variable A causally affects a variable B if and only if interventions on A are correlated with variations in B, where an intervention on a variable A is a cause of A's taking the value it does that has no effect on any effect of A except via its effect on A. It may be true that in Newcomb's Problem, the fact that A_1 and A_2 are potential *actions* of yours implies that your intention, act of will or some other psychic state or event counts as an intervention on the variable that reflects which of A_1 and A_2 obtains. But it is part of the stipulated arrangement in Newcomb's Problem that S_1 and S_2 *are*

4 Restriction on Opportunities

You might think that a fair comparison of the track records of one-boxing
and two-boxing should compare, not the return for those options simpliciter, but the return for each in cases that match over *the available
opportunities*. The reason that this is fair is that we are supposed to be
settling what it is best for you to *do* in the instance of Newcomb's Problem
that you now face; so, we ought to be comparing the outcomes that you *can*
bring about in *this* situation. But if you can realize (if it is in your present
power to realize) a situation in which you get K by two-boxing, then you
cannot realize a situation in which you get M by one-boxing: one-boxing
can only get you nothing. More generally, for you now, the practically
relevant comparison of two options that are currently available is not
between the overall track record of each: it is between their track records
in cases where the available outcomes are the same as those that are
available to you now.

By this measure two-boxing clearly outperforms one-boxing. In your actual
situation, either (i) the available alternatives are: getting M by choosing A_1
and getting $M + K$ by choosing A_2 (if the prediction was that you will choose
A_1); or (ii) the available alternatives are: getting nothing by choosing A_1 and
getting K by choosing A_2. Either way, in situations where the available
alternatives are the same as yours now, A_2 certainly does better than A_1. That
is why it is rational to choose A_2.

This argument is presumably what lies behind the idea that those who
choose A_2 and end up with K are not irrational but just unlucky. They could
have been dealt a better hand; but they could not have played a better game.
Joyce imagines a dialogue between Irene, who has just taken A_1 and ended up
with M, and Rachel, who has just chosen A_2 and ended up with K. Rachel asks
Irene why she didn't choose A_2, which would have gotten her an extra K.

> If Irene really does [understand Newcomb's Problem], then she can give no
> good answer to Rachel's question since, having gotten the M, she must
> believe that she would have gotten it whatever she did, and thus that she
> would have done better had she [chosen A_2]. So, while she may feel

---

correlated with the choice between values of that variable, i.e., with your intervention upon it; so
again, the partition $\{S_1, S_2\}$ fails to be causally independent of your choice, and the argument
against choosing A_1 fails to get off the ground. (The chapter by Stern in this volume has further
discussion of what interventionism implies about Newcomb's Problem.)

superior to Rachel for having won the million, Irene must admit that her choice was not a wise one when compared to *her own* alternatives.[11]

In the same vein, if Irene asks Rachel: "Why Ain'cha Rich?," her reply is that being rich was never an available outcome. "We were never given any choice about whether to have a million. When we made our choices, there were no millions to be had."[12]

4.1 C-opportunity and E-opportunity

Let us address this by asking whether there is a sense of "opportunity" in which it is both true and of practical relevance that (i) Irene had an opportunity to get $M + K$ and (ii) Rachel had no opportunity to get M. There certainly is a relatively familiar sense of "opportunity" in which both claims are true. Let us say that an agent has a *C-opportunity* to get a prize X at a time t if either (a) the agent does get X because of something that she does at t or (b) if the agent had chosen to act in some other way at t the agent would have got, or would have had a non-negligible chance of getting, X. The counterfactual in (b) is meant to be interpreted in terms of *causal* independence: what an agent would have got, had she chosen otherwise, is what she does get in some possible situation that matches the actual situation over all matters of fact that are causally independent of her choice, except possibly for some local miracle just before her choice.

Clearly Irene had a C-opportunity to get $M + K$. It was true, causally independently of her choice, that she had been predicted to one-box. In any possible situation that holds that fact fixed and where she two-boxes, she gets $M + K$. So if she *had* two-boxed, she would have gotten a prize of $M + K$. Similarly, Rachel lacked any C-opportunity to get M. It was causally independently of *her* choice that she had been predicted to *two*-box. In any possible situation that holds that fact fixed and where she chooses to one-box, she gets nothing.

But we are also familiar with a kind of opportunity that exists where the C-opportunity to get it does not, and which can be present when the

---

[11] Joyce 1999: 153. The first interpolation glosses Joyce's supposition that Irene believes the counterfactuals that capture the causal structure of Newcomb's Problem. In our terms, these are: (i) if in fact she was predicted to choose A_1, she would have gotten at least M whatever she had done; (ii) if in fact she was predicted to choose A_2, she would have gotten at most K whatever she had done.

[12] Lewis 1981b: 38.

corresponding C-opportunity is absent. Consider a lottery that works as follows. Long ago, the organizers set up a pseudorandom process by means of a computer algorithm that every few seconds spits out a number between 1 and 100 million. The sequence of numbers is entirely determined by the initial state of the machine, but all such numbers come up equally often in the long run, and there is practically no way of calculating the next number at any point in the sequence. Today, everyone is invited to enter the lottery for a fee; and everyone who does enter is assigned a number between 1 and 100 million that is determined by his or her genetic profile. The big prize is awarded to the first person whose number comes up after noon tomorrow. Nobody would have any difficulty in saying that today, *all* entrants have the same opportunity to win the big prize. But only *one* entrant has any C-opportunity to win the big prize. What everyone has at that point is a kind of opportunity that is not a C-opportunity.

In the other direction, consider a – slightly fanciful but perfectly possible – game of "Rock, paper, scissors" played in a Libet-style environment, involving fMRI scanners and the like, in which your opponent can very reliably predict your choice in any game several seconds before you make it; she makes *her* choice in the light of the prediction. (If you are predicted to play "paper," she plays "scissors," and so on.) Anyone who plays this game against this prescient opponent will certainly lose, and it is perfectly natural to say about such a game, after losing, that the predictive apparatus deprived you of any opportunity to win ("You never stood a chance"). But at the time of your decision you *do* have a C-opportunity to win: what the opponent chooses in the game is causally independent of your choice at the time of decision, and a world in which the opponent's choice remains fixed but yours is appropriately different is one at which you *do* win. What you lack in this game of "Rock, paper, scissors" is certainly some kind of opportunity to win, but it isn't the C-opportunity to win, which you possess as fully in this version of the game as you do in any regular case.[13]

It would take more space than I have here to extract from these and similar examples a precise single notion of opportunity. But as a first pass, we might start by identifying a notion of opportunity that is relevant to the epistemic perspective of the deliberating agent. Let us say that a proposition

---

[13] This example comes from Dennett (1984: 204 n. 16). The lottery example modifies a case on which he writes: "Would many people *buy* a ticket in a lottery in which the winning stub, sealed in a special envelope, was known to be deposited in a bank vault at the outset? I suspect that most ordinary people would be untroubled by such an arrangement, and would consider themselves to have a real opportunity to win" (1984: 132).

represents an *E-opportunity* for a deliberating agent if her confidence in its truth is not independent of her current intentions. For example: suppose that you are wondering whether to enter the lottery. The proposition that your genetic profile is the winning profile is *not* an E-opportunity for you at that point, since your confidence in its truth remains fixed at a presumably very low level p, however strongly you are inclined to enter the lottery. But the proposition that you will win the big prize *does* count as an E-opportunity for you at that time (as it does for everyone eligible): as you become more inclined or less inclined to enter it, your confidence that you will win the big prize will vary between zero and p. Similarly, suppose that you are wondering what to play in the Libet-style game. You have E-opportunities to play rock, paper or scissors; but you lack any E-opportunity to win, since your confidence that you'll win is not only low, but independent of what you currently intend to play.

Deprivation of E-opportunities can occur whilst one's C-opportunities remain fixed; and there are ways in which this can happen that the agent will reasonably regard as a restriction of her liberty. If one is knowingly in the grip of an insidious kind of addiction, one may deliberate whether to indulge, knowing all the time that one will eventually talk oneself into (say) smoking the cigarette. "Talking oneself into" doing something is certainly a kind of deliberation, but if you know in advance what its upshot will be, you lack the E-opportunity to do anything else. By contrast, this addict has (and may know that he has) the same C-opportunity not to smoke as anyone else. Again, suppose that I have implanted a microchip in your brain that influences the course of your deliberation in such a way that given any (C-)opportunity to smoke, you will always end up choosing *not* to smoke. On learning of this you might reasonably complain that I have deprived you of a kind of liberty worth caring about, but I haven't deprived you of any C-opportunity (and you may know this). This is not to say that there is an objective kind of liberty of which the addiction or the implantation themselves have genuinely deprived you; rather, the position is that the agent will regard these sorts of interference or control *as* restrictions of her liberty, because learning that they obtain itself contracts her E-opportunities.[14]

---

[14] There is clearly a great deal more to say on this point than space permits. For instance, the present account of E-opportunities implies that you do not have an E-opportunity to win at a regular game of "Rock, paper, scissors" if you are equally confident in all of your opponent's possible plays. For further discussion of the connection between ignorance of the upshot of one's deliberation and one's sense of oneself as a free agent, see my 2014a sections 8.1–2.

In short: (i) there are two different kinds of opportunity, E-opportunities and C-opportunities; (ii) in some cases where they diverge, it is the E-opportunities that capture what we intuitively regard as the real opportunities; and (iii) in other cases, it is E-opportunities, and not either C-opportunities or his beliefs about them, that co-vary with the agent's sense of having the kind of liberty that we seem to care about.

4.2 Application to the Newcomb Case

Now return to the Newcomb situation. Clearly in that case, your C-opportunities for gain are *either*: to get M and to get $M + K$, *or*: to get nothing and to get K. Your C-opportunities are to get M and to get $M + K$ if the state is S_1, that is, if you were predicted to choose A_1; your C-opportunities are to get nothing and to get K if the state is S_2, that is, if you were predicted to choose A_2.

What about your E-opportunities in the Newcomb case? There are two kinds of case. If you are certain that the predictor has got you right this time, then you will be certain that you will get either M or K. Your confidence that you'll get M will rise or fall as your intentions incline more strongly or less strongly towards A_1. Similarly, your confidence that you will get K will rise or fall as your intentions incline more strongly or less strongly towards A_2. So, your E-opportunities are the proposition that you get M and the proposition that you get K. But your getting $M + K$, and your getting nothing, are not among your E-opportunities. You lack these E-opportunities because you are certain that they will not obtain, and you remain certain of this however your intentions vary.

On the other hand, if you are highly confident of the predictor's powers, but not certain of them, then you have more E-opportunities. Suppose that you are throughout your deliberations 90% confident that the predictor has got you right. Then as you incline towards A_1 you become more confident that you will get M, and more confident that you will get nothing; as you incline towards A_2 you become more confident that you will get $M + K$, and more confident that you will get K. So in this second type of case, all four outcomes in Table 1 are among your E-opportunities.

But the important points to take from either case are as follows. First, your E-opportunities in a Newcomb Problem are *never* the same as your C-opportunities in it: it is never the case that your E-opportunities are exactly to get M and to get $M + K$, and it is never the case that your E-opportunities are exactly to get nothing and to get K. Second, if we look at the statistical return to one-boxing and two-boxing across a population of Newcomb agents whose *E-opportunities* match your own, we see that the average return for one-boxing exceeds that for two-boxing in either case. Across a population in which everyone's E-opportunities are to get M and to get K (as when the predictor is knowably perfect), the average return for A_1 is M and the average return for A_2 is K. Across a population in which everyone has an E-opportunity to get any of the four notionally possible prizes, the average return for A_1 is close to M and the average return for A_2 is close to K (how close will depend on the predictor's accuracy).[15] Either way, the return for one-boxing exceeds the return for two-boxing.

The result of all this is to cast a new light on the claim that A_2 does better than A_1 across a population of Newcomb-type agents with *the same opportunities* as you. The claim is false if by "opportunity" we mean E-opportunities, though it is true if by "opportunities" we mean C-opportunities. Moreover, it looks at least plausible that the comparison of options *can* reasonably be made on the basis of E-opportunities. Consider again a deterministic lottery of a type like that described earlier, and suppose I am wondering whether to enter tomorrow's lottery. Wouldn't it make a decisive difference to my deliberations to know the average return per entry across the *whole* population? If this average exceeds zero, then I will enter; if it does not, then I will not. But that just means knowing the return across people whose *E-opportunities* are the same as mine. By contrast, I cannot know the average return for people with my *C-opportunities* for gain, since I have no idea what my C-opportunities are. (Either everyone with my C-opportunities who enters will win the big prize, or everyone with my C-opportunities who enters will win nothing.) It therefore seems that in this case, the practically relevant statistic holds fixed my E-opportunities, not my C-opportunities. "Why Ain'cha Rich?" simply applies parallel reasoning to the Newcomb Problem, and it concludes that one-boxing is plainly the better choice.

4.3 E-opportunity vs. Evidence of C-opportunity

The gist of the argument in this section so far is that it is equality of E-opportunity, not equality of C-opportunity, that matters when comparing

---

[15] More precisely: if we write Fr for the probability distribution that matches frequencies amongst all trials, the expected return for A_1 is $ER(A_1) = MFr(S_1|A_1)$ and the expected return for A_2 is $ER(A_2) = (M + K)Fr(S_1|A_2) + KFr(S_2|A_2)$. The first quantity exceeds the second if and only if $Fr(S_1|A_1) - Fr(S_1|A_2) > K/M$, so it suffices for one-boxing to do better that S_1 obtain only slightly more often when you choose A_1 than when you choose A_2.

the average return to contemplated options in a decision problem, and that it follows from this that such a comparison does after all vindicate one-boxing in the Newcomb Problem.

One possible response to this is that I have confused E-opportunity, which needn't matter, with *evidence* of my C-opportunity, which does. To see this, return to the lottery example. A statistic governing the whole population in the lottery case is practically relevant – you might say – not because it tracks my E-opportunity, but because it *tells* me something about my C-opportunities. Suppose that of those who enter amongst those who share my E-opportunities (i.e., the entire population), a proportion p wins a big prize, and the remaining $1 - p$ win nothing. Then I should have a confidence of around p that I have a *C-opportunity* to win a big prize, and a confidence of around $1 - p$ that I don't. As p rises, I will eventually become confident enough that anyone with my *C-opportunities* does better by entering than by not entering, and at that point I will, if I am rational, enter the lottery myself.

The point of this response is that it rationalizes the decision to enter the lottery, given that enough people have won it in the past, without also rationalizing one-boxing in the Newcomb Problem. To see this, note that what ultimately rationalizes the decision to enter the lottery, according to this story, is my being confident enough that people with my C-opportunities who enter the lottery will do better than people with my C-opportunities who do not. But *your* confidence in the corresponding claim in the Newcomb Problem is always zero. You know all along that *either* (i) you have the C-opportunity to get M by one-boxing and the C-opportunity to get $M + K$ by two-boxing, *or* (ii) you have the C-opportunity to get nothing by one-boxing and the C-opportunity to get K by two-boxing. Learning about the fortunes of other Newcomb agents may affect your *relative* confidence in these two hypotheses, but it has no effect on your certainty that *either* (i) or (ii) is true. And since both (i) and (ii) entail that the average return to two-boxing exceeds the average return for one-boxing amongst people with the same C-opportunities as you, it follows that learning about the fortunes of other agents in the Newcomb Problem cannot rationalize your one-boxing there, even though learning about the fortunes of other lottery entrants *can* rationalize entering the lottery in cases where doing so is intuitively plausible.

But that response has counterintuitive consequences elsewhere. Consider a Libet-type scenario in which you are playing a "matching" game with very many choices. Specifically, suppose that you must choose to press one of 101 buttons. If you press any of the buttons numbered 1–100, you win a prize if and only if your opponent, who has the same choices, presses a *different* button from you. If you press any of these buttons and your opponent presses the same button, you *lose* the value of the prize. If you press the last button, labeled X, nothing happens (Table 2). Your opponent doesn't see what button you press, and in fact her choice is causally independent of yours, but she *is* able to predict your choice a few seconds before you make it, by means of an fMRI scanner. You know all of this.

Table 2: Matching Game (MG)

	Predicts A_1	Predicts A_2	...	Predicts A_{100}	Predicts A_{101}
A_1: press 1	-1	1	...	1	1
A_2: press 2	1	-1	...	1	1
...	...	...	...	...	...
A_{100}: press 100	1	1	...	-1	1
A_{101}: press X	0	0	0	0	0

Because the opponent is so good at predicting, the only time anybody avoids a loss in MG is by pressing X. Thousands of people have played this game, and anyone who presses any other button always loses $1. This illustrates the fact that amongst people with your E-opportunities in MG, those who press any button other than X have an average loss of $1, while those who press button X have an average loss of exactly zero. But this statistic does not tell you anything about your C-opportunities. If you are indifferent about pressing any of the buttons 1–100, you should be close to certain that *amongst people with your C-opportunities*, those who press button 1 do better than those who choose button X, and you should be close to certain that those who press button 2 do better than those who choose button X, and so on.

It is worth going over why. Suppose that shortly before choosing you are almost certain that the opponent will correctly predict your move and equally confident that you will press (because indifferent about pressing) any of the buttons 1–100. This means that your confidence that the opponent will predict that you press button 1 is no more than 1%. So you are at least 99% confident that anyone with your C-opportunities who does press button 1 will get $1, and of course you know that anyone with your C-opportunities (or anyone else) who presses X will get nothing. So you are at least 99% confident that amongst people with your C-opportunities, those who press button 1 do better than those who press button X.

If the practical relevance of statistical information runs via its effect on your confidence about the fortunes of people with your C-opportunities, as the objection says, then in MG it is rationally permissible to press button 1 (and equally so to press button 2, button 3,… button 100.) But this is highly counterintuitive. The intuitive response to this problem is that pressing button X is the only rational thing to do.[16] And this intuition again aligns with the statistics connected with E-opportunity: you know in advance – whatever your credence about what you will in fact do – that people with your *E-opportunities* who press button X do better (because they lose nothing) than people with your E-opportunities who press any other button (because they always lose $1).

The response therefore fails. Consideration of various cases suggested that when we assess any one of your options by its statistically expected return, we should be aggregating across agents who share your E-opportunities, not across agents who share your C-opportunities. The objection was that doing so is practically relevant when, but only when, the statistic provides *evidence* bearing on the relative expected return for each option amongst people who share your *C-opportunities*; since these relative returns are already known in the Newcomb Problem, it is irrelevant that in the Newcomb Problem one-boxing generates a higher return overall. But that objection is implausible because of its consequences for the case that I just considered.

5 Conclusion

We have looked at four possible responses to Why Ain'cha Rich? None of them appears to supply any independently motivated grounds for rejecting the argument. Of course, this is very far from saying that WAR should move any consistent defender of Causal Decision Theory. Such a person could I think quite reasonably cite the third and fourth responses as *her* reasons

---

[16] Of course, one can say things that make it seem more rational to press button 1. After all, since you are practically certain that the opponent does not predict that you press button 1, you are practically certain that you would have done better to press button 1 than to press button X. And you are practically certain that if you were to press button X, then you would regret it: you would know (when all is revealed) that if you had pressed button 1, then you would have made another $1. Still, none of this stops it from being true that if these considerations move you to press button 1, then you will lose $1.

for not finding the argument compelling. But I don't think anyone who starts with an open mind should take those responses to supply such reasons. "As surely as water will wet us, as surely as fire will burn," so too will one-boxing enrich us, and two-boxing will certainly not; and these plain statements of fact are surely of practical consequence.

4   Epistemic Time Bias in Newcomb's Problem*

Melissa Fusco

1 Introduction

Suppose you will either experience an hour of pain on Tuesday and an hour of pleasure on Thursday, or an hour of pleasure on Tuesday and an hour of pain on Thursday. Neither possibility seems clearly better than the other – unless, that is, today is Wednesday. For a fixed menu of pleasures and pains, *time-biased agents* prefer distributions wherein more pains are in the past, and more pleasures are in the future.

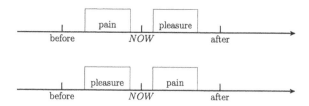

Figure 1: Basic time bias.

The two possibilities in Figure 1 illustrate this basic commitment.

A classic discussion of time bias can be found in Chapter 8 of Parfit's *Reasons and Persons* (Parfit 1984).[1] Parfit imagines that he wakes up in the hospital, and is told by a nurse that *either* he has just survived a very long and painful operation (which he does not remember, since patients undergoing his surgery are regularly given postoperative medication which induces amnesia), *or* he has yet to undergo a shorter version of an operation of the same type. While Parfit's character generally prefers to suffer less, rather than more, he now finds that he would much prefer to learn that he has already had the longer, earlier operation. His

---

* Warm thanks to Arif Ahmed, Lara Buchak, Alex Kocurek, and John MacFarlane (via pandoc) for their help on this chapter.

[1] Parfit distinguishes two forms of bias, *future* bias – being more concerned about one's future welfare than for one's past welfare – and *near* bias – being more concerned for the near future than for the distant future. Future bias is thought to be more rationally defensible than near bias (see *op. cit.*: 158 ff.): I focus only on future bias here.

preferences seem to have been reversed by the mere passage of time. Nor is this a reversal between outcomes which are intuitively equal. Rather, Parfit's character now prefers a possibility in which his life contains *more* total hours of suffering.

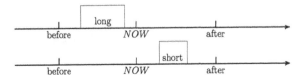

Figure 2: Parfit's surgeries.

In this chapter I investigate a form of time bias which is entailed by a common form of causal decision theory. It can be seen in the causalist's response to Newcomb's Problem.

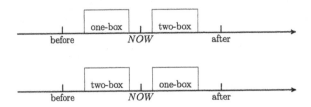

Figure 3: Newcomb time bias.

A standard causal decision theorist in Newcomb's Problem – an agent who prefers to two-box, given the chance – will nonetheless prefer to learn that Newcomb Problems which involve her, but over which she currently has *no agency*, are ones in which she one-boxed. When Parfitian amnesia is on the scene, time bias easily takes root: learning about one's past actions is a paradigm of being invested in a decision over which one does not currently exercise agency.[2] Hence an agent, located at *NOW*, who sees the past as fixed but the future as hers to choose, will prefer the upper state in Figure 3 to the lower one. As we will see, the CDTer's time bias is, if anything, stranger than the time bias investigated by Parfit, because it is traceable to a normative, rather than affective, source.

One puzzle about this setup will immediately present itself. I say that, in Figures 1–3, time-biased agents prefer the lower possibility into the upper.

---

[2] Arif Ahmed points out to me that an act's *being in the past* is not *essential* to the difference between choosing and learning. The relevant type of CDTer will also prefer to learn that she *will* one-box in the future. This is true, but such "learnings" are puzzling because of the difficulty many people (myself included) will have reconciling such foreknowledge with free agency. For this reason, I will focus in this chapter on learning about the past.

But – it will be objected – an agent located at *NOW* cannot make such a switch, since that would involve changing her past. This is true, but I show in Section 3 that the basic structure of the time-biased preferences illustrated in Figure 3 can, with suitable manipulation, be turned into actionable preferences which operationalize the basic dispute between the biased and unbiased perspectives.

Nonetheless, this is a confessional chapter: I *maintain* the CDTer's commitments in both cases, despite the (literal) costs. Damage control commences in Section 4. I will argue that *Evidential* Decision Theory is also committed to a form of time bias. The question, then, is whether the two forms of bias have an equal claim to be features of rational behavior.

2 Evidence and Decision

In this section I sketch the shared background of causal and Evidential Decision Theory with an eye to two issues, screening-off and conditionalization, which bear on the argument to come.

I will assume that in all decision problems, agents have the goal of maximizing expected utility. They come to these problems equipped with a value function $Val(\cdot)$ on possible worlds, which can be lifted onto propositions called *outcomes*;[3] following standard idealization, I will occasionally speak as if the units in the range of the value function are equivalent to dollars. An agent's decision problem at a time t can be represented as a triple $\langle \mathcal{A}, \mathcal{S}, Cr^t \rangle$, where

- $\mathcal{A} = \{a_1, \ldots, a_n\}$ is a set of *acts* available to the agent at t,
- $\mathcal{S} = \{s_1, \ldots, s_m\}$ is a set of admissible *states of nature*, where a set of states is admissible in a decision problem only if each act-state conjunction $(a_i \wedge s_j)$ is assigned a determinate image $Val(a_i \wedge s_j)$ under the value function, and
- $Cr^t(\cdot)$ is the agent's time-t credence function.[4]

At a first pass, the agent's credence function $Cr^t(\cdot)$ will be conceived of in Bayesian terms. I assume that it is a probability function, and that the agent is

---

[3] The lifting of the value function can be achieved as follows. Where $Val(\cdot)$ is a worldly value function in $W \to \mathbb{R}$, (i) $Val(\{w\}) = val(w)$, and (ii) $Val(p \cup q)$ is defined only if $Val(p) = Val(q)$, in which case $Val(p \cup q) = Val(p)$.

[4] In what follows, I treat the members of $\mathcal{A}$, $\mathcal{S}$, and the domain of $Cr^t(\cdot)$ alike as *propositions*. This differs from Savage's original picture of the relevant primitives, on which acts are *functions* from states to outcomes. See Joyce 1999: Ch. 2 for discussion of this shift.

disposed to respond to new information by conditioning this function on the strongest proposition of which she is informed. I'll also assume that her expectations are generally *frequentist*, in the sense that when they concern events in the present or future, they will, *ceteris paribus*, conform to observed past frequencies over event classes of the same type.

Against this background, we can introduce Savage (1972)'s decision theory. It applies the general statistical notion of expectation to the value of an act $a \in \mathcal{A}$.[5]

> **Equation 1** (Savage Expected Utility). $SEU^t(a) = \sum_i Cr^t(s_i) Val(a \wedge s_i)$.
>
> **Savage's norm**: maximize expected utility at t by choosing an act $a \in \mathcal{A}$ such that $SEU^t(a)$ is maximal: that is, such that $SEU^t(a) \geq SEU^t(a')$, for any $a' \in \mathcal{A}$.

EDT and CDT can then be framed as two different ways of cashing out a final condition Savage put on well-formulated decision problems: the requirement that the set S of states in the problem are, at time t, *independent* of the agent's acts.

EDT construes act-state independence in terms of probabilistic confirmation. In general, the EDTer holds that rationality requires calculating expected utilities in terms of act-conditioned probabilities $Cr^t(s|a)$:

> **Equation 2** (Evidential Expected Utility). $EEU^t(a) = \sum_i Cr^t(s_i|a) Val(a \wedge s_i)$.
>
> **Conditional Probability norm**: maximize expected utility by choosing an act $a \in \mathcal{A}$ such that $EEU^t(a) \geq EEU^t(a')$, for any $a' \in \mathcal{A}$.

This does justice to one attractive gloss on Savage's independence requirement for S: it entails that Savage's equation will apply to any problem in which $Cr^t(s|a)$ is equal to $Cr^t(s)$ – in other words, to any problem in which the agent takes a to be *probabilistically* independent of s.

The CDTer, on the other hand, favors an interpretation of act-state independence that is *causal*. She rejects the EDTer's gloss because probabilistic independence – a lack of probabilistic (dis)confirmation – is neither necessary, nor sufficient, for causal independence.[6] In this chapter, I will use Lewis

---

[5] Where F is a function, the generic notion of expected value says that $E[F] = \sum_i f_i Pr(F = f_i)$. Here, our probability function is the subjective credence function $Cr^t(\cdot)$, and the function F is $Val(\cdot)$ across act-state pairs, where the act is held fixed.

[6] For non-necessity, we can consider any case (such as Newcomb) in which an act a probabilistically confirms some state s whose truth-value was fixed before a occurred. For non-sufficiency, we can appeal to examples such as Hesslow (1976)'s well-known thrombosis case: while birth control pills cause thrombosis, the effect is statistically masked because birth control pills counteract pregnancy, which *itself* causes thrombosis.

(1981a)'s "dependency hypothesis" characterization of the CDT. This takes as primitive a partition $\mathcal{K}$ of dependency hypotheses, which are *stipulated* to be causally independent of the agent's available acts, as well as sufficient to fix an outcome value for each $(a \wedge k_i)$ conjunction.[7]

> **Equation 3** (Causal Expected Utility). $CEU^t(a) = \sum_i Cr^t(k_i)Val(a \wedge k_i)$, where $\mathcal{K} = \{k_1 \ldots, k_n\}$ is a partition of dependency hypotheses.

> **Causal Support norm:** maximize expected utility by choosing an act $a \in \mathcal{A}$ such that $CEU^t(a) \geq CEU^t(a')$, for any $a' \in \mathcal{A}$.

This way of stating CDT leaves open the model-theoretic implementation of what is special about the k's – leaves open, that is, how doxastic states might be modeled so as to *reflect* an agent's conviction that the k's are causally independent of acts in $\mathcal{A}$. It is clear, however, that these modeling requirements will exceed those of austere Bayesianism.[8] Note that, since EEU is partition-invariant, the evidential decision theorist can also calculate expected utility using the $\mathcal{K}$-partition.[9]

We are now in a position to define *NC problems*, decision problems with the general structure of Newcomb's puzzle. For my purposes, an NC problem will be a problem with two acts $\mathcal{A} = \{a^*, \neg a^*\}$ and two causal dependency hypotheses $\mathcal{K} = \{k_1, k_2\}$, in which a dominance argument across $\mathcal{K}$ gives the agent a one-utile incentive to choose a^*; nonetheless, because $\neg a^*$ indicates a much more valuable outcome (which I'll refer to generically as "the big prize"), the conditional probability norm favored by the EDTer recommends $\neg a^*$. Hence from the CDTer's point of view, the distinctive thing to say about NC problems is that maximizing expected utility will entail getting some "bad news on the side" (Joyce 2007: 542) – news that statistically disconfirms one's getting the big prize.

Here is an example that we will revisit throughout.

> **Mood Candles.** Lighting aromatic mood candles ($= L$) slightly increases happiness. Bob is deciding whether to light a mood candle. He also finds himself unable to

---

[7] For a probabilistic version of the same idea – which does not require that dependency hypotheses *entail* outcomes, but merely that they fix their chances – see §10 of Lewis 1981a.

[8] Candidates include the directed acyclic graphs of Pearl 2000, and the worldly distance metrics of Lewis (1973)'s treatment of natural language counterfactuals.

[9] Evidential expected utility is *partition invariant* in the sense that the expected utility assigned to all acts $a \in \mathcal{A}$ w.r.t. a credal state $Cr^t(\cdot)$ will be the same in any problems $\langle \mathcal{A}, \mathcal{S}_1, Ct^t(\cdot) \rangle$ and $\langle \mathcal{A}, \mathcal{S}_2, Ct^t(\cdot) \rangle$ so long as $\cup \mathcal{S}_1 = \cup \mathcal{S}_2$. See Joyce 1999: 176 ff., and references therein, for discussion of this feature.

remember whether he suffers from depression (= D). In his current state of ignorance and indecision, his possible outcomes, and their values, are:

Table 1

	Not Depressed ($\neg D$)	Depressed (D)
Don't light ($\neg L$)	9	0
Light (L)	10	1

There's a catch, though: knowledge of the therapeutic value of mood candles is rare. Only people who are depressed tend to have it, since only they are regularly updated by their doctors about mood therapies. Hence Bob's credences are such that $Cr^t(D|L) > Cr^t(D)$: while lighting mood candles modestly conduces to happiness, it is a strong indicator of *unhappiness*.

A flat-footed (perhaps *naive*; see below) calculation by the EDTer's conditional probability norm will recommend that Bob refrain from lighting a mood candle, so long as L raises the statistical probability of D by more than $1/9$.[10] Given that Bob also believes that lighting mood candles cannot *cause* depression, however, it is unclear that this is the right recommendation.

2.1 Screening Off

The formal description of an NC problem is clearly coherent, and *Mood Candles* prima facie fits the bill. Some philosophers have argued, however, that there *are* no NC problems in real life – at least, not for rational agents. This is a common factor of two venerable threads in the literature: the

---

[10] Calculation:

$$EEU^t(\neg L) \qquad > EEU^t(L) \text{ iff}$$
$$\sum_i Cr^t(k_i|\neg L)\,Val(k_i \wedge \neg L) \quad > \sum_i Cr^t(k_i|L)\,Val(k_i \wedge L) \text{ iff}$$
$$> [Cr^t(\neg D|L) \times 10 + Cr^t(D|L) \times 1] \text{ iff}$$
$$> [Cr^t(\neg D|L) \times 10 + Cr^t(D|L)]$$

Since $Cr^t_\phi(\cdot) = Cr^t(\cdot|\phi)$ is itself a probability function, this is equivalent to

$$[Cr^t_{\neg D} \times 9] \quad > [Cr^t_{\neg D} \times 10 + (1 - Cr^t_{\neg D})]$$

Letting $n = Cr^t_{\neg D}$ and $m = Cr^t_{\neg D}$, this obtains just in case

$$9n \quad > 10m + (1 - m) \text{ iff}$$
$$n \quad > m + 1/9.$$

so-called "Ramsey thesis" (Ahmed 2014a: Ch. 8; Ramsey 1926 [1990]), and Ellery Eells's "tickle defense" of EDT (Eells 1982). The shared idea is that as long as a responsible agent takes her full pre-decision evidence into account, and is not irrationally influenced by factors outside of her control, she will be able to access evidence Z that *screens off* the statistical support that her act would apparently provide for any state k:[11]

> **Screening off.** Z screens off k from a relative to a probability function $Cr^t(\cdot)$ just in case $Cr^t(k|a \wedge Z) = Cr^t(k|Z)$.
>
> **A screening problem.** An agent's decision problem $\langle \mathcal{A}, \mathcal{K}, Cr^t(\cdot) \rangle$ is a screening problem just in case the agent's evidential position at t makes accessible some Z which screens off all $a \in \mathcal{A}$ from all $k \in \mathcal{K}$, relative to $Cr^t(\cdot)$.

When some such Z is available pre-decision, the additional evidential value of actually acting – actually making a instead of $\neg a$ the case, or vice versa – is nothing. $Cr^t(k) = Cr_Z^t(k) = Cr_Z^t(k|a)$ for all a and k, so the characteristic decision rules of EDT and CDT will always endorse the same act.

Mood Candles is designed to suggest screening-off. There are two particularly appealing candidates for the proposition Z which would screen off L from D. The first will likely have occurred to the reader up above: it comes from Bob's ability to reflect on what he already believes about the problem he is facing. In *Mood Candles*, Bob believes that lighting up gives him an extra utile, and that it is causally independent from whether he gets the big prize. (As noted above, exactly how his doxastic state reflects this conviction about causal independence is a question I am currently leaving unanswered, but he *does* believe it.) It follows that Bob already has evidence that he is a member of a certain class: the class of people who are aware that lighting mood candles is efficacious for happiness. This is statistical evidence for D – by the description of the problem, only depressed people tend to be that informed about (small, but) effective ways of elevating one's mood. The first candidate "screener" proposition, then, supports a high credence in D. On this interpretation, Bob's mere middling credence in D looks rather like a bit of epistemic negligence.[12]

---

[11] Eells's discussion emphasizes the first idea – screening evidence is always available pre-decision – while Ramsey emphasizes the second – that an agent not irrationally influenced by factors outside of her control in fact *cannot* learn anything from her acts. Beyond Eells, the issue is discussed in Ahmed 2014a: Ch. 8 and Briggs 2010.

[12] In the Newcomb literature, this thought finds an ally in the argument that if an agent can reflect that her "decision algorithm" recommends two-boxing, she can already conclude that the predictor left no money in the opaque box. For a version of this argument, see Yudkowsky 2010: 26 ff.

A second candidate is also introspective, but pulls in the other direction. It focuses on Bob's introspective access to how he feels, rather than what he believes. *D*, after all, is the proposition that *Bob is depressed at t*. Surely agents have some privileged access to whether they are depressed, which comes from concentrating on how they feel. By the description of the problem, though, Bob doesn't feel strongly depressed.[13] Reflection on this supports a *low* credence in *D*. On *this* interpretation, Bob's mere middling credence in *D* might still be remiss – but because it should be *lower* than it is, rather than higher.

Of course, there is not really an "either-or" in this case: Bob should be epistemically responsible in taking into account *both* what he (already) believes and how he (already) feels. Bob may be in such a singular evidential situation vis-à-vis his total evidence that it is just difficult to say what a justified credence in *D* would be.

This is why the suggestion that real-life NC problems are all screening problems has had an equivocal effect on the literature. Briggs weighs in on the thesis as follows:

> Whether the tickle defense rules out all cases of conflict between EDT [and] CDT ... [is] controversial. Perhaps agents can be reasonable enough to look to decision theory for advice without being as rational and self-aware as [screening-off] requires ... [However,] it seems clear that a *great many* suitable agents will have information that renders their actions evidentially irrelevant to the dependency hypotheses. To the extent that situations [to the contrary] are rare, the task of deciding among various possible decision theories is rendered much less urgent. (Briggs 2010: 28, emphasis added)

Two nondenominational morals, I think, emerge from considerations of the screening-off debate. The first is that screening-off arguments in NC problems will always support the dominant option – that is, the act prima facie favored by the CDTer, rather than the EDTer.[14] The second is that an agent's pre-decision reflection on her full information – an activity I will call

---

[13] I construe this to be underlying the fact that $Cr^t(D)$ isn't high, in the original description of the case.

[14] Why? In a 2×2 NC problem, let $\mathcal{A} = \{a^*, \neg a^*\}$ and $\mathcal{K} = \{k_1, k_2\}$. By screening off, there is some Z which the agent can update on before acting such that $\forall k : Cr^t_Z(k|a) = Cr^t_Z(k)$. It follows from this that $Cr^t_Z(k|a^*) = Cr^t_Z(k|\neg a^*)$ for all $k \in \mathcal{K}$. Calling this number n, we have $EEU(a^*) = nVal(a^* \wedge k_1) + (1 - n)Val(a^* \wedge k_2)$, and $EEU(\neg a^*) = nVal(\neg a^* \wedge k_1) + (1 - n)Val(\neg a^* \wedge k_2)$. By dominance, $Val(a^* \wedge k_1) > Val(\neg a^* \wedge k_1)$ and $Val(a^* \wedge k_2) > Val(\neg a^* \wedge k_2)$. Hence, $EEU(a^*) > EEU(\neg a^*)$. This is not an embarrassment to the EDTer's theory, since screening off makes going for dominant options compatible with that theory.

epistemic entrenchment – must be considered by *both* theories in any putative NC problem, even if full screening-off is controversial. This often leads to subtle disputes about what introspection can deliver, from the agent's point of view.

For CDTers, entrenchment serves as a reminder that causalists do not actually ignore their evidence. In NC problems, dominance reasoning can obscure the fact that even a CDT-rational agent must have some *particular* credence over each dependency hypothesis in $\mathcal{K}$ before she acts. Decision procedure aside, she will not make the best decisions she could make if these credences fail to reflect her total evidence.[15]

EDTers, for their part, need epistemic entrenchment to combat the objection that their view recommends patently absurd acts that would apparently result from giving too much weight to general statistical correlations. Consider Pearl's complaint:[16]

> [According to EDT,] [p]atients should avoid going to the doctor to reduce the probability that one is seriously ill; workers should never hurry to work, to reduce the probability of having overslept ... remedial actions should be banished lest they increase the probability that a remedy is indeed needed. (Pearl 2000: 108–9)

The reasonable reply here, on behalf of the evidentialist, appeals to an agent's need to incorporate her full information into her credal state before she acts. Focus on Pearl's example of hurrying and lateness (which we can call *Hurrying to Class*). Grant Pearl's assumption, that hurrying (e.g., to a 10am class) on any given day is correlated with lateness on that day:

Table 2: Hurrying to Class. $Cr^t(\text{late} \mid \text{hurry}) > Cr^t(\text{late})$

	Not Late ($\neg L$)	Late (L)
hurry (H)	9	0
don't hurry ($\neg H$)	10	1

---

[15] For a similar observation about how dominance arguments can obscure a CDTer's evidential duties, see Joyce (2012: 239). Here I take the traditional view that epistemic norms such as the Principle of Total Evidence are *sui generis* principles of epistemic rationality, whose justification is prior to the justification of any practical maxims they inform (see, e.g., Hare and Hedden 2016: 615). For an argument that inverts this order of priority – offering a practical vindication of the Principle of Total Evidence – see Good (1967); for an argument that EDT, in contrast to CDT, has trouble underwriting Good's Theorem, see Skyrms (1990a).

[16] I take the quote from Pearl via Ahmed (2014a: 82).

Why doesn't the EDTer's norm recommend that I avoid hurrying today, right now, as I head to campus?

In the simplest case, I have a timekeeping device with me, and either do, or easily could, glance at it, to see that it is, for example, 9:46am. Relative to any specific time – *a fortiori*, relative to 9:46am – the correlation between lateness and hurrying is reversed:

$$Cr^t(\text{late} \mid \text{hurry} \wedge 9{:}46\text{am}) < Cr^t(\text{late} \mid 9{:}46\text{am}).$$

Hence, relative to my full information, hurrying is *not* contraindicated by my act-conditionalized credences.

What happens if I do not have a watch? It seems that the same reasoning ought to hold – after all, forbearing to hurry when one is unsure of the time seems like an even *worse* idea than forbearing to hurry when one *does* know the time.

Here is a way to generalize the strategy. The argument sketched above clearly did not depend on the time's being exactly 9:46. Rather, it depended on the fact that the agent (here, me) was located at *some* particular time, which she could think about in a direct way. Suppose that agents generally have the power to pick the current times rigidly by means of some demonstrative token – say, μ.[17] Relative to my full information in *Hurrying to Class*, I accept

$$Cr^t(\text{late} \mid \text{hurry} \wedge \text{the time is } \mu) < Cr^t(\text{late} \mid \text{the time is } \mu).$$

Just in virtue, then, of the patently introspectable fact that the time is μ, an EDT agent in a problem like *Hurrying to Class* can conclude that there is no in-respect-of-lateness reason not to hurry. We will return to this argument in Section 4.

2.2 Conditionalization and Lewis's Package

Up above, I described credence by saying that agents are presumed to have Bayesian dispositions – including a disposition to update by conditionalizing on any new evidence they receive. This presentation, however, was neutral with respect to the normative status of that tendency. Bayesianism's additional, normative claim is that conditional probabilities provide a *diachronic norm* of belief revision – that an agent *ought* to revise her credence in light of

---

[17] See Moss (2012)'s "Dr. Demonstrative" example (§1) for a similar maneuver, in the case of self-locating beliefs.

evidence E by moving from $Cr^t(\cdot)$ to $Cr^{t+}(\cdot) = Cr^t(\cdot|E)$. I will call this norm *Conditionalization* (with a capital "C.")

Conditionalization entails, as a special case:

(CNC) In an NC problem, one ought to update by conditionalizing on one's chosen act a.

(CNC) appears to dovetail well with the CDTer's conciliatory position on news values. Relative to a fixed credal state, propositions like L – *a mood candle is lit* – or H – *the agent hurries* – carry a stable news value both before, and after, that proposition is an available act. The CDT position is simply that, when, for example, L *is* an available option, it should be *chosen* (or passed up) according to its causal expected utility rather than its news value.

Lewis himself seems to have embraced (CNC).[18] The conjunction of (CNC) and the CDTer's expected utility equation comprise a joint CDTer position which we can call "Lewis's package" (LP):

(LP) At all times t, one ought to:

(i) assign expected utility to an available act a using one's time-t credences across the dependency hypotheses $\mathcal{K}$, selecting the *CEU*-maximal act;

(ii) when making propositions in $\mathcal{A}$ true by one's actions, plan to evolve $Cr^t(\cdot)$ according to Conditionalization.

Conceptually, there is something odd about (LP). If an agent with Lewis's form of CDT is self-aware, she can *anticipate* what her credence in the various dependency hypotheses $k \in \mathcal{K}$ will be at any future time t^+ when she has performed a given available act a.[19] By Conditionalization, these future credences are just

$$\{Cr^t(k|a) : k \in \mathcal{K}\}$$

By Equation 2, these are the same numbers the *EDTer* will *currently* apply to calculate the expected value of a. In favoring *CEU* over *EEU* as a guide to action, then, the Lewisian CDTer *de facto* holds that one's current credences

---

[18] In Lewis (1976b: 302), he writes that "there are good reasons why the change of belief that results from coming to know an item of new evidence should take place by conditionalizing on what was learned." A footnote from the quoted sentence refers the reader to Teller (1973)'s Dutch Book argument in favor of Conditionalization. Lewis advances his own Dutch Book for Conditionalization in Lewis (1999).

[19] And has learned nothing more: a is, at t^+, her *total* evidence.

in the dependency hypotheses overrule the credences one plans to adopt. This seems in tension with the idea that there is something more informed about the more opinionated credence function $Cr^t(\cdot|a)$, *in virtue of which* it is the one an agent should plan (following Conditionalization) to transition to upon performing *a*.

This tension can be seen in the juxtaposition of norms governing action and belief change. For the Lewisian CDTer, a lack of causal efficacy makes it the case that one should, in choosing an act, be *unmoved*, so to speak, by the fact that the act provides statistical evidence for a desirable state. To return to our first example, in light of its lack of efficacy in promoting *D*, Bob should be *unmoved*, in his decision-making, by the fact that *L* provides good statistical evidence for *D*. But because the Lewisian CDTer also thinks Conditionalization applies, he believes that if Bob *does* bring about *L*, he must, in a way, be "moved" by its evidential force after all: Bob is rationally required to increase his confidence in *D* upon processing the fact that he made *L* true.[20]

3 The Newcomb Hospital

Can this be right? The present section will be devoted to sketching an odd consequence of (LP), drawing on an analogy with Parfit's amnesia cases. The setting is a hospital funded by an eccentric billionaire, where amnesiac patients – beset by various handicaps – get to face daily Newcomb puzzles.

Suppose you wake up in the Newcomb hospital today, knowing that you faced Newcomb's Problem yesterday, but unable to remember what you did. You are completely evidentially indifferent on the matter, assigning a probability 1/2 to the proposition that you one-boxed and a probability 1/2 to the proposition that you two-boxed. The news that you did the former would be a near-to-perfect indication that you made $1000 yesterday, and the news that you did the latter would be a near-to-perfect indication that you made $1M yesterday. Hence, in your state of total ignorance, you estimate your winnings at $500,500. A Lewisian rational amnesiac, while a two-boxer, much prefers the news that she one-boxed in the past.

A trade reflecting this preference can be engineered.

---

[20] Again, Lewis seems to have been comfortable with this result. Addressing Newcomb, he writes with an air of resignation that a two-boxer must expect one-boxers to come out ahead: "[w]e [two-boxers] ... did not plead surprise. We knew what to expect" (Lewis 1981b: 378). See Byrne and Hájek (1997) for more discussion of Lewis's package view.

Newcomb Past. Fry (a CDTer) awakens, bedridden and with amnesia, in the Newcomb Hospital. Today is Wednesday, and Fry is in bed 336. Bender, Fry's hospital roommate, in bed 335. While Fry can't remember whether he one-boxed on Tuesday, Bender truthfully tells him that *he*, Bender, one-boxed on Tuesday. Fry's winnings from yesterday are in a lockbox marked "Tuesday-336." Bender's are in a lockbox marked "Tuesday-335." For Δ, Bender offers to switch lockbox keys with Fry (call this trade σ).

What is the value of σ? Because Fry conditionalizes on his evidence, when Bender tells him that he one-boxed yesterday, Fry becomes 90% confident that Bender's lockbox, Tuesday-336, has $1M$ in it. He estimates the value of his *own* lockbox at a mere $500,500. So, σ, the deal Bender is offering, is attractive unless the amount of money he demands for the trade, Δ, is greater than $399,500.

The attractiveness of the trade in *Newcomb Past* reflects the fact that, on Lewis's picture, it is a very fine thing to have the past of a one-boxer. Of course, it is also a fine thing to *be* a two-boxer, since two-boxing maximizes causal expected utility. We can easily engineer a bet whose appeal reflects the latter preference as well, by inducing some uncertainty over which act is being performed:

Newcomb Present. It's still Wednesday in the hospital with daily Newcomb rounds. But now it's time for Fry to choose today's move. It works like this: the predictor has already deposited either $1M$ or nothing in opaque lockbox "Wednesday-336," depending on whether she predicted that Fry would, today, choose act 2B (which dumps an extra thousand into Wednesday-336). Likewise, she has already deposited either $1M$ or nothing in Bender's lockbox, "Wednesday-335," depending on whether she predicted Bender would perform 2B (which dumps an extra thousand into Wednesday-335). Unfortunately, Fry has an injury that makes speech impossible, and due to a mix-up, the nurse incorrectly believes he is a monolingual Dingbats speaker. The nurse gives him the day's Newcomb menu with two options printed entirely in Dingbats: he can either choose "@" or "#." He does not know which one corresponds to 2B. In frustration, Fry randomly circles the first option, "@." Bender's form is in English – he can choose either "1B" or "2B" – and it is still blank. For Δ, Bender offers to switch forms with Fry (call this trade τ).

How should Fry evaluate τ? The value of the trade depends on what he would do with Bender's form if he got it. But that part is easy: from the Causal Point of View, a two-box form in a Newcomb Problem is always worth $1000 more than a one-box form (it ensures that the extra thousand is dumped into one's box). So, a CDTer like Fry, who chooses τ and gets a blank form will certainly go on to pick the two-box option. Indeed, the trade Bender is offering is appealing to him so long as $\Delta < \$500$.

Argument: let β be the amount of money already in the box. As in the original Newcomb puzzle, because the identity of β is causally independent of anything Fry does now, we can frame the problem from the CDTer's point of view via an argument that abstracts from the value of β:

Table 3

	$p\,(@ = 1B)$	$\neg p\,(@ = 2B)$
τ	$\beta + (1000\text{-}\Delta)$	$\beta + (1000\text{-}\Delta)$
$\neg\tau$	$\beta + 0$	$\beta + 1000$

If Fry is completely indifferent on the identity of @, then $Cr(p) = .5$. Hence

$$
\begin{aligned}
CEU^t(\tau) \quad &= .5(\beta + (1000 - \Delta)) + (1 - .5)(\beta + (1000 - \Delta)) \\
&= \beta + (1000 - \Delta), \\
CEU^t(\neg\tau) \quad &= .5(\beta + 0) + (1 - .5)(\beta + 1000) \\
&= \beta + 500.
\end{aligned}
$$

Hence $CEU^t(\tau) > CEU^t(\neg\tau)$ iff $\Delta < \$500$. (For a presentation of the argument without the variable β, see the Appendix.)

3.1 Putting It All Together

Our last step to getting the preferences reflected in Figure 3 is to combine the betting behavior of the CDTers in *Newcomb Past* and *Newcomb Present*. The following setup, while baroque, will do the job:

Newcomb Hospital Fusion. As before, it's Wednesday. The lockboxes, Fry's amnesia, and mutual knowledge that Bender one-boxed yesterday are as in *Newcomb Past*. As in *Newcomb Present*, Fry has been given a Dingbats form for today's round and has randomly marked "@" on it, while Bender was given a form in English. Bender offers Fry a "combo deal": for n, he will switch lockbox keys *and* forms with Fry (call this trade ω).

Before Fry makes his decision, he learns one additional piece of information from the Dingbats-language chart by his bed: yesterday, the act he chose was #.

How will Fry weigh the combo deal? The causal expected utility of choosing "@" on this round is equal to the causal expected utility of choosing "#" given his current credences, since each act has a 50% chance of being the act of one-boxing and a 50% chance of being the act of two-boxing. Because he knows he chose # yesterday and has currently marked @ on his form, he knows, even before he trades, that he is in one of the two situations in Figure 3. For an additional Δ – by choosing ω – he can guarantee the outcome a Lewisian CDTer confidently desires: that he has a one-box-*past* and two-box-*future* as opposed to the reverse.

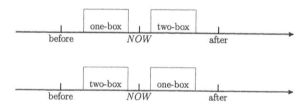

Newcomb time bias: Figure 3, repeated.

As in *Newcomb Present*, the part of the combo deal that ensures a two-boxing present is appealing. Even *more* appealing is the part that mimics being able to ensure a one-boxing past – that is, Fry's being able to trade his Tuesday box for Bender's. Since Bender one-boxed yesterday, his box, Tuesday-335, is 90% certain to contain a million dollars. Indeed, as long as $\Delta < \$400{,}500$, Fry will choose to trade.

Argument: let β be the amount of money that is already in Fry's *Wednesday* box – that is, the box he will get today. No matter what, Fry is not trading away his Wednesday box (rather, if he takes ω he switches *Tuesday* boxes and Wednesday *forms* with Bender.) So the value of β – though it is act-dependent – is, again, causally independent from what Fry chooses to do.

If Fry chooses ω, he switches forms with Bender, so he insulates his own payoff from the identity of @; he will get $\beta + (1000 - \Delta)$ today in any case. Fry's payoffs will then further depend on whether Bender's Tuesday box contains $\$1M$ or nothing (let these be propositions $5T1$ and $5T0$, respectively).

If Fry declines the combo deal, opting for $\neg\omega$, he keeps the contents of his own Tuesday box, and his payoffs will then depend on whether his Tuesday box contains $\$1M$ or nothing (let these be propositions $6T1$ and $6T0$, respectively). Moreover, his total payoff does now depend on the identity of @ – he will get an extra thousand just in case @ $= 2B$ (row 2, below).

Table 4: Utilities in the 8-state decision matrix in Newcomb Hospital Fusion. Again, $\neg$ is the proposition that @ $= 1$-boxing.

	I.$5T1, p, 6T1$	II.$5T1, p, 6T0$	III.$5T1, \neg p, 6T1$	IV.$5T1, \neg p, 6T0$
ω	$1M + (1000 - \Delta) + \beta$	$1M + (1000 - \Delta) + \beta$	$1M + (1000 - \Delta) + \beta$	$1M + (1000 - \Delta) + \beta$
$\neg\omega$	$1M + 0 + \beta$	$0M + 0 + \beta$	$1M + 1000 + \beta$	$0M + 1000 + \beta$
	V.$5T0, p, 6T1$	VI.$5T0, p, 6T0$	VII.$5T0, \neg p, 6T1$	VIII.$5T0, \neg p, 6T0$
ω	$0M + (1000 - \Delta) + \beta$	$0M + (1000 - \Delta) + \beta$	$0M + (1000 - \Delta) + \beta$	$0M + (1000 - \Delta) + \beta$
$\neg\omega$	$1M + 0 + \beta$	$0M + 0 + \beta$	$1M + 1000 + \beta$	$0M + 1000 + \beta$

That yields:

Table 5: Credences in Newcomb Hospital Fusion.

	I.$5T1, p, 6T1$	II.$5T1, p, 6T0$	III.$5T1, \neg p, 6T1$	IV.$5T1, \neg p, 6T0$
ω	0.045	0.405	0.405	0.045
$\neg\omega$	0.045	0.405	0.405	0.045
	V.$5T0, p, 6T1$	VI.$5T0, p, 6T0$	VII.$5T0, \neg p, 6T1$	VIII.$5T0, \neg p, 6T0$
ω	0.005	0.045	0.045	0.005
$\neg\omega$	0.005	0.045	0.045	0.005

What of the appropriate causalist credences? Knowing Bender one-boxed yesterday, Fry is 90% confident in $5T1$, the proposition that the predictor put $\$1M$ in Bender's box on Tuesday. Hence he has credence 90% across columns I-IV and credence 10% across columns V-VIII. Within each of these possibilities, he is 90% confident the predictor predicted his own move (#) correctly yesterday, so he expects a correlation of .9 between $\neg p$ (which entails that # is one-boxing) and $6T1$, the proposition that the predictor put $\$1M$ in Tuesday-336. Finally, his overall confidence in p is .5. A calculation shows that $CEU(\omega) = 900,000 + (1000 - \Delta)$ over the baseline β, while $CEU(\neg\omega) = \$500,500$ in excess of the same baseline. Hence Fry will take the deal as long as $\Delta < \$400,500$.

3.2 Karma Foretold

From the Lewisian point of view, time-biased trading at the Newcomb Hospital seems to be motivated. Of course, the causal decision theorist wants the *past* of a one-boxer – this is an excellent indicator that she is already a millionaire! And of course, she wants to two-box *today* – that guarantees that she walks away with $1000 more than she would have gotten otherwise![21]

These sentiments are articulated from the point of view of a particular time – for example, the time labeled "*NOW*" in Figure 3. But at the point in time marked "after" – when both rounds lie in the past – taking the combo deal will seem, by the CDTer's own lights, like bad news. More to the point, I fear, it looks like an *irrational* decision. Suppose Fry wakes up amnesiac on *Thursday*, and is told that Bender offered him the combo deal on Wednesday. On Wednesday, Fry could have kept his @ form for free, ensuring that the action he performed on Wednesday was different in kind from the action he performed on Tuesday – hence, that he was in *one* of the scenarios in Figure 3, without knowing which. Now, for Fry to learn whether he took Bender's combo deal is to learn either that:

(a) In the last two days, he two-boxed once and one-boxed once, in some unknown order, or
(b) In the last two days, he one-boxed once and two-boxed once, and willingly gave Bender $400,000 to ensure that the first happened before the second.

Though I am a CDTer, the idea of learning that *I* secured (b) over (a) in such a situation makes me uncomfortable. It is hard to justify the claim that one way of ordering my acts in time is worth (more than) $400,000 more than the other.[22]

---

[21] I here use the counterfactual locution Lewis favored in describing causalist reasoning; see, e.g., Lewis 1981b.

[22] It is worth noting that, while the post-factum news value of (a) exceeds that of (b), *Newcomb Hospital Fusion* is not a Dutch Book: choosing (b) over (a) does not *guarantee* that you are poorer in every possible world (though it does make it likely). A world w will make the agent better off, where:

- Tuesday-335 contains $0 in w,
- Tuesday-336 contains $1 million in w, and
- @ is one-boxing.

For a diachronic Newcomb Problem for the two-boxer that *does* make a standard CDTer poorer in every possible world, and the extra assumptions this involves, see Ahmed's "Newcomb Insurance" cases (Ahmed 2014a: 202ff., 2017).

4 CDT's Response

The oddity in *Newcomb Hospital Fusion* arises because the tie between one-boxing and being a millionaire is invisible to Fry's deliberations at the moment of choice – only to reappear with a vengeance later, when he assesses what he believes is in yesterday's lockboxes. This makes it impossible to stably assume that Fry does, or doesn't, take the fact that an agent is a one-boxer to be *good reason* to increase one's confidence that that agent is a millionaire.

In most ordinary situations, there does seem to be a stable fact. In *Mood Candles*, for example, the causalist's intuition is that Bob should light a candle even though candle-lighting is – at least, in the general population – correlated with depression. It also seems like, if he does light up, he *fails* to acquire a good reason to increase his confidence that he is depressed. At least, that is what I suggested above: there is something odd about thinking, with Lewis, that if Bob does light up, he would thereby acquire good evidence for thinking he was depressed after all. He *knows* why he's contemplating lighting up – he is in pursuit of the extra utile, which the act will secure him – and it's not *because* he is depressed.

4.1 Time Bias and EDT

As indicated in the introduction, my first defensive maneuver is to argue that EDT is *also* susceptible to time bias. It will be up to the reader to determine whether the EDTer's form of bias is as unflattering as Fry's.

Recall that in the Pearl-inspired example *Hurrying to Class* (§2.1), hurrying on any given day was generally correlated with lateness on that day:

$$Cr^t(\text{late} \,|\, \text{hurry}) > Cr^t(\text{late})$$

Nonetheless, relative to a specific time – like 9:46am – that correlation failed (in fact, was reversed), so that relative to the agent's full available information hurrying was *not* after all contraindicated, by either decision theory. For cases where the agent lacked a wristwatch or other timekeeping device, I extended the screening-off argument by appeal to introspection – to the agent's ability to "lock on" to her current time by means of some demonstrative μ, and use $Z = $ *the time is μ* to screen off the support lent to lateness by hurrying.

This ability, which extends only to the present moment, gives rise to time bias. Suppose an EDTer, who has just woken at t unable to remember what she has done for the last month (the interval of time "T^-"), finds herself in an iterated version of *Hurrying to Class*. If she cares about her track record of arriving on time to class, she may well:

(i) prefer to hurry at t, and
(ii) prefer to learn that she *didn't* hurry at any past time $t' \in T^-$.

The EDTer in *Hurrying to Class* should have attitude (i) for the reasons explained above: overturning Pearl's argument depends on her leveraging her (difficult-to-express) knowledge *that it is now the current time*. But this power is limited to t: with regard to her past, it is rational to continue to maintain attitude (ii), *despite* taking up attitude (i). Though something similar to the thought involving μ was true in the past – after all, for any morning in the agent's past *it was that time then* as much as *it is this time now* – the proposition expressed by the former thought does not have the force, at t, that the latter thought does. EDT's appeal to screening off in many putative NC problems will commit the view to time bias in cases where, as in *Newcomb Hospital*, those NC problems are iterated.

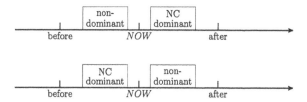

Figure 4: A shared structure (top preferred to bottom).

It is worth thinking about why this is so, and what this means for Lewis's views on updating. At issue in Lewis's package is whether an agent – for example, in an NC problem – should *plan* to conditionalize on the act she performs. One way of thinking about this is to ask whether, in an NC problem, your attitude towards the acts $a' \in \mathcal{A}$ you are currently contemplating should be the same as the preferences you would have if you woke up tomorrow with amnesia and were about to be *told* which $a' \in \mathcal{A}$ you performed.

But introspection's power to create and discover propositions concerning the time t of choice is sharply different across the two situations. In *Hurrying to Class*, I can think to myself, *the time is μ*. Oddly, at the time of choice, I seem to gain some knowledge by this exercise – knowledge that affects my rational calculations. Later, post-amnesia, I will not have this power. It is patently uninformative to be told, "you reflected *then* that the time was ν, where by ν you meant: the time it in fact was."

A comparison between this situation and a CDTer's reasoning in NC problems is telling. In an NC problem, I can tell myself, "I will now choose the dominant option because of the (small) benefit I get from it." I am introspecting my motives, which give me a reason to perform a^* *without* giving me an extra reason to think that I won't get the big prize. If I am subsequently subjected to amnesia, later I will only be able to learn that I either performed the dominant option or I didn't. Any extra knowledge I had access to at the earlier time will be wiped out: it feels uninformative to be told "you performed the dominant option because it was the dominant option." While not *vacuous*, this thought seems, again, *obvious* – like the thought that if you hurried, you hurried at some particular time.

5 Conclusion

I began by arguing that CDT is time-biased, and by going on to show that this leads to odd choices in the Newcomb Hospital. In response to this bias, I sketched a *tu quoque* against EDT. The dialectical purpose of the sketch is clear enough: time bias may be a pervasive feature of decision-making, not a weakness in any particular theory of it.

It would be best to close by returning to Newcomb itself. EDTers may feel that, in shifting the focus to an agent's act-prior viewpoint, and the epistemic reasons that can be articulated from it, I have quietly sidestepped the real bite of the original Newcomb Problem. For in that puzzle, whether it is possible, *before you act*, to come to an evidentially supported credence in the $\mathcal{K}$-partition {*Predictor predicted you'd one-box, Predictor predicted you'd two-box*} (={$P1, P2$}) is precisely what is at issue. There is a widespread tendency in the literature to assume

> (P) In Newcomb's puzzle, the only evidence available with respect to distributing credence over the partition {$P1, P2$} is how the player him/herself reasons about the expected utility of one-boxing.

This would apparently render full reflection, and any attempt at screening off, impossible. Ahmed, for example, presses the point that "[a]nyone facing Newcomb's Problem has *no* evidence that relevantly distinguishes him now from anyone else whom the statistical generalization [$\Pr(P1|1B)$ and $\Pr(P2|2B)$ are high] covers; that is, all other persons who ever face this problem."[23] This suggests (P)'s epistemological moral – that there is just no

---

[23] Ahmed 2014a: 191.

way, in Newcomb, to take your full evidence into account *before you act*, since *nothing* counts as relevant evidence other than the act itself.

But while (P) might characterize *Newcomb's Problem*, it *isn't* supported, either by deductive or abductive considerations, by the basic mechanics of an NC problem – where the latter is described simply in terms of causal dependency hypotheses, credences, and outcome states, and where it is the latter that usefully diagnoses the split between Causal and Evidential Decision Theory.

Appendix: *Newcomb Present*: Long Argument

As in the original Newcomb Problem, Fry's estimate of how much money is already in his own box – the quantity called "β" in the main text – is act-dependent. It depends, in part, on how likely he thinks it is that he *will* take the trade. Fry knows that if he chooses to trade, then he will certainly two-box. The predictor will have predicted this with her usual accuracy, so that $Cr^t(P1|\text{trade}) = .1$ and $Cr^t(P2|\text{trade}) = .9$. However, if Fry declines to trade, sticking with the wildcard move @, then the probability that the predictor predicted $P1$ is equal to the probability that the predictor predicted $P2$ (the high correlations of $P1$ with $1B$ and $P2$ with $2B$ are now balanced by the fact that Fry has a 50% chance of doing $1B$ and a 50% chance of doing $2B$).

$$\mathcal{K} = \{(P1 \wedge p), (P1 \wedge \neg p), (P2 \wedge p), (P2 \wedge \neg p)\}$$

is a dependency hypothesis partition.

Table 6: Utilities.

	$P1 \wedge p$	$P1 \wedge \neg p$	$P2 \wedge p$	$P2 \wedge \neg p$
τ	$1M + (1000 - \Delta)$	$1M + (1000 - \Delta)$	$0M + (1000 - \Delta)$	$0M + (1000 - \Delta)$
$\neg \tau$	$1M + 0$	$1M + 1000$	$0M + 0$	$0M + 1000$

We have:

$$
\begin{aligned}
Cr(k_1) = Cr(P1 \wedge p) &= Cr(p)Cr_p(P1) \\
&= Cr(p)\left[Cr_p(P1|\tau)Cr_p(\tau) + Cr_p(P1|\neg\tau)Cr_p(\neg\tau)\right] \\
&= Cr(p)\left[.1Cr_p(\tau) + .9Cr_p(\neg\tau)\right] \\
&= .5\left[.1Cr_p(\tau) + .9Cr_p(\neg\tau)\right] \\
&= .5\left[.1Cr(\tau) + .9Cr(\neg\tau)\right]
\end{aligned}
$$

$$
\begin{aligned}
Cr(k_2) = Cr(P1 \wedge \neg p) \\
= Cr(\neg p)Cr_{\neg p}(P1) \\
= Cr(\neg p)\left[Cr_{\neg p}(P1|\tau)Cr_{\neg p}(\tau) \right. \\
\left. + Cr_{\neg p}(P1|\neg\tau)Cr_{\neg p}(\neg\tau)\right] \\
= Cr(p)\left[.1Cr_{\neg p}(\tau) + .1Cr_{\neg p}(\neg\tau)\right] \\
= .5\left[.1Cr(\tau) + .1Cr(\neg\tau)\right] \\
= .5(.1) = .05,
\end{aligned}
$$

$$
\begin{aligned}
Cr(k_3) = Cr(P2 \wedge p) \;\; &= Cr(p)Cr_p(P2) \\
&= Cr(p)\big[Cr_p(P2|\tau)Cr_p(\tau) + Cr_p(P2|\neg\tau)Cr_p(\neg\tau)\big] \\
&= Cr(p)\big[.9Cr_p(\tau) + .1Cr_p(\neg\tau)\big] \\
&= .5\big[.9Cr_p(\tau) + .1Cr_p(\neg\tau)\big] \\
&= .5\big[.9Cr(\tau) + .1Cr(\neg\tau)\big]
\end{aligned}
$$

$$
\begin{aligned}
Cr(k_4) &= Cr(P2 \wedge \neg p) \\
&= Cr(\neg p)Cr_{\neg p}(P2) \\
&= Cr(\neg p)\big[Cr_{\neg p}(P2|\tau)Cr_{\neg p}(\tau) + Cr_{\neg p}(P2|\neg\tau)Cr_{\neg p}(\neg\tau)\big] \\
&= Cr(p)\big[.9Cr_{\neg p}(\tau) + .9Cr_{\neg p}(\neg\tau)\big] \\
&= .5\big[.9Cr(\tau) + .9Cr(\neg\tau)\big] \\
&= .5(.9) = .45.
\end{aligned}
$$

For shorthand, we stipulate that $t = Cr(\tau)$. Hence Fry's credences are as follows:

Table 7: Credences, take I.

k1	k2	k3	k4
.45 − .4t	.05	.4t + .05	.45

Letting $z = .4t$, this is equivalent to:

Table 8: Credences, take II.

k1	k2	k3	k4
.45 − z	.05	z + .05	.45

Calculating causal expected utilities:

$$
\begin{aligned}
CEU^t(\tau) \;\; &= (.45 - z)(1M + (1000 - \Delta)) + .05(1M + (1000 - \Delta)) \\
&\quad + (z + .05)(0M + (1000 - \Delta)) + .45(0M + (1000 - \Delta)) \\
&= (.5 - z)(1M) + (1000 - \Delta), \\
CEU^t(\neg\tau) &= (.45 - z)(1M + 0) + .05(1M + 1000) + (z + .05)(0M + 0) \\
&\quad + .45(0M + 1000) \\
&= (.45 - z)(1M) + (.05)(1M) + (.05)(1000) + (.45)(1000) \\
&= (.45 - z + .05)(1M) + (.05 + .45)(1000) \\
&= (.5 - z)(1M) + 500.
\end{aligned}
$$

Once again, we have something of the form: $CEU^t(\tau) = X + (1000 - \Delta)$, while $CEU^t(\neg\tau) = X + 500$. Hence $CEU^t(\tau) > CEU^t(\neg\tau)$ iff $\Delta < 500$. QED.

5   Newcomb's Problem is Everyone's Problem: Making Political and Economic Decisions when Behavior is Interdependent

Robert Grafstein

A decision-maker facing Newcomb's Problem is supposed to assume that her own decisions do not causally affect the financial decisions of the supernatural entity she encounters however much their decisions are stochastically entwined. She is supposed to assume, of course, that she can causally affect her own behavior. What is this natural entity that affects her behavior? How does such an entity bring this about?

These lead-off questions are not in fact the ones I propose to answer in this chapter. Rather, I will argue, the attempt to answer these questions, implicitly or explicitly, represents a conflation of normative and (social) scientific analysis. The entity at issue – the decision-making self – is supposed to project itself above the stochastic entanglements described in Newcomb's Problem to make decisions that escape the expected utility implications of its decisions. The decision-making self exercises free will.[1]

In this context, the scientific status of free will or free choice is analogous to that of God. It's fair to say that, as a rule, scientists who are theists are also compartmentalizers. However deep their theism, they don't inject God into their scientific explanations. In other words, they don't commit the mistake made by the scientist in the famous Sidney Harris cartoon who inserts into his mathematical analysis "then a miracle occurs."

How theistic scientists reconcile this compartmentalization with what is presumably a non-compartmentalized world is their business. For my purposes the important point is that at least as a matter of legitimate scientific style, scientific explanation precludes theism. Since I don't have an additional 2500 years to debate the ontological status of free will, I propose to invoke the

---

[1] Thus, Meek and Glymour (1994) consider the standard example of a decision-maker who is genetically disposed to contract lung cancer and to smoke, although smoking is assumed not to cause lung cancer. They posit a "Will" that intervenes by deciding whether to smoke, thereby negating the genetic cause of smoking. I am grateful to Arif Ahmed for pointing me to this analysis.

same stylistic compartmentalization in the present context: free will, I will assume, is not a scientifically legitimate explanatory variable.

Let me say that by excluding free will as an explanatory variable I mean to exclude a positive role for it. In my book it's perfectly fine to adopt the language of free will or free choice as a shorthand way of saying that the decision-maker wasn't subject to waterboarding, under hypnosis, under the influence of a psychotropic drug, and so on. Free choice, in this case, indicates normal conditions. These normal conditions allow the decision-maker to make a rational choice. There is sufficient time, freedom from distractions, and freedom from overriding or contravening psychological or physical influences to make a decision that rationally links decisions to the decision-maker's preferences and beliefs.

Rational choice is where the normative and scientific converge (or collide): rationality can and has been interpreted both ways. Although controversial, I assume throughout that individuals behave rationally. Not because they choose to be rational, but just as an empirical fact about their behavior. The normative side of rational choice focuses on a different question: what should decision-makers do should they choose to behave rationally? There is no normative force to the answer – no effective advice – if decision-makers must behave rationally simply because of the way they are.

The obvious problem with adopting a scientific rational choice approach in the context of Newcomb's Problem is the resulting need to deal with the supernatural or certainly implausible entity with which decision-makers interact. Undertaking a social scientific analysis of fictions does not seem very promising. Therefore, I shift terrain to the one-shot Prisoners' Dilemma game (PD). The parallel between Newcomb's Problem and PD has often been noted and often disputed (see, e.g., Lewis 1979; Walker 2014; Bermúdez 2015a, b). For my purposes, the link is the following. The conditional expected utility (CEU) of choosing one box in the case of Newcomb's Problem is higher than the CEU of choosing two.[2] This expected utility calculation is conditioned on all available information *and* the action in question. Comparison with PD works only when the probability of the second player's doing action A (i.e., cooperate or defect) is higher given the first player's doing A than given the first player's doing $\sim A$. In both Newcomb's Problem and PD, of course, we stipulate that the decision-maker's choices have no causal impact on the other party (the supernatural entity and the second player respectively). In this

---

[2] CEU theory is associated with Jeffrey (1983), although he explores a number of refinements to avoid the implications of his approach for Newcomb's Problem.

sense, Newcomb's Problem and PD are connected, but the latter involves a social interaction subject to social scientific analysis.

Our task, then, is to determine what a rational decision-maker (strictly speaking, a rational game player) does – not should do – when confronted with a PD in which the relevant inequality in conditional probabilities is satisfied. Of course, this task still seems to be burdened by a residual normativity. Even if we are not interested in advising a decision-maker about the best choice to make, we still must determine which behavior is rational, and this would seem to mean determining which behavior is best in light of the decision-maker's preferences and beliefs. In this setting "best" seems normative: an action may be best because it maximizes CEU, best because it implements a minmax strategy, best because it is rationally admissible (Levi 2012), and so on.

From a social scientific standpoint, "best" means best empirical explanation, considering not only empirical accuracy but also parsimony (to ward off cognitive psychology's growing list of "rules" governing human cognition). I will set that issue aside in favor of the main event: CEU maximization versus causal decision theory (CDT).[3]

1 Empirical Issues

While we do not have any empirical experience with a naturally occurring Newcomb's Problem, behavior in PD situations has been extensively examined. Empirical findings vary greatly and are highly sensitive to nuances of experimental research design (see, e.g., Ledyard 1995). But a fair summary is that cooperation is higher than one would expect according to standard (causal) game theory, since in the PD, cooperative behavior is strictly dominated by noncooperative behavior.

The empirical problem facing CDT is that, strictly speaking, any degree of cooperation is inexplicable within the theory. In contrast, CEU maximization can explain both cooperation and defection depending on the relative payoffs from both and on the perceived probabilities of cooperation and defection given cooperation and defection respectively. Mass democratic two-party elections offer a less abstract illustration of CEU's applicability. Since Downs (1957), standard rational choice theory has faced the challenge of trying to explain turnout in this very important case. The challenge arises because the

---

[3] It is worth noting that these are not the only two alternatives; see, e.g., Levi (1975) and Wedgwood (2013).

probability of a given voter's single vote causing one of the candidates to win – the probability of a tie vote without that single vote – is negligible, so the expected benefits associated with a preferred candidate's winning are swamped by the positive costs of participation. If so, there apparently is mass irrationality in mass elections. This so-called paradox of voter turnout is "the paradox that ate rational choice theory" (Fiorina 1990: 334).

Ledyard (1984), however, notes that this decision theoretic analysis is misleading: if rationality dictates abstention and this same decision is duplicated across the electorate, then abstention is no longer rational. Any voter could tip the election. Of course, a one-voter outcome is not an equilibrium, nor is participation across the board, which remains irrational. Instead, depending on the size of the threshold cost of voting, there will be varying but positive turnout from a rational electorate. Unfortunately for rational choice theory, Ledyard provided the standard theory only a temporary respite. Palfrey and Rosenthal (1985) demonstrated that when turnout has a positive cost it limits to zero as the size of the electorate increases. Evidently a positive turnout in this case falls short of a realistically positive turnout.

Yet CEU maximization can bridge the gulf. Observe that each of a candidate's supporters plays a PD with the others. In its simplest form, the PD logic dictates that whatever the turnout of other supporters, there are too many who find that abstention dominates participation. As a consequence, turnout collapses. CEU maximization, however, can undermine this empirically unreasonable conclusion. Suppose supporters identify with other supporters due to shared ethnicity, race, geography, values, or party identification. This identification can affect conditional expected utility insofar as the probability of turnout from a supporter conditional on the voter's own is higher than the unconditional probability that informs the causal decision-maker's calculations. Under these circumstances, the CEU of participation can exceed the CEU of abstention even in mass elections (see Grafstein 1991, 1999). These stochastic dependencies will still vary depending on the depth and breadth of group identifications and the extent to which they cut across candidates. But substantial turnout is now a theoretically possible outcome.[4]

The socio-political identities anchoring this empirical defense of CEU theory are not merely impressionistic or anecdotal. Quattrone and Tversky

---

[4] There is an opposite problem for this analysis. Except under knife-edge conditions, all voters with a reasonable cost of voting will participate (Grafstein 1991: 1001–1002). The way out of the problem of excessive participation is to assume, plausibly, that voters are not certain about the stochastic dependency parameters, but in Bayesian style have probability distributions characterizing their beliefs about them.

(1988) document the operative force of what they condescendingly call the "voters' illusion." Similarly, in a large number of experiments, cognitive psychologists have shown the impact of a "false consensus effect," the tendency of experimental subjects to infer the characteristics of others from their own, characteristics that are here taken to include their own behavior (see, e.g., Fiske and Taylor 1984; on Bayesian principles, Dawes 1990 argues for the potential non-falsity of this effect). To summarize the overall empirical lesson, "One of the most replicated findings from social dilemma experimental research is of a strong positive relation between expectations of others' cooperation and one's own cooperation," where the "expectations often occur *subsequent* to one's own choice" (Orbell et al. 1991: 61; emphasis in the original).

For CDT, of course, the question is whether this realism is purchased at the cost of the rationality assumption. This is the burden of residual normativity that we must address. Nevertheless, we will not address it entirely on the usual philosophical terrain of CEU's normative claims against CDT's normative claims. As we will see, the fact that we are analyzing PD social scientifically, that is, without recourse to a causally empowered self that possesses free will, changes the nature of these normative claims.

2 Scientific Issues

CEU maximization is the best policy for decision-makers because it maximizes their well-being by their lights. Specifically, when the level of stochastic dependence in a PD involving players 1 and 2 is sufficiently high, we have the following set of stochastic facts:[5]

$$\mathrm{Prob}(\mathrm{Cooperate}_2|\mathrm{Cooperate}_1) > \mathrm{Prob}(\mathrm{Defect}_2|\mathrm{Cooperate}_1), \quad (1)$$

---

[5] Technically, we treat states of the world and acts homogeneously as propositions. The resulting beliefs and actions are modeled in terms of the Boolean algebra on those propositions. See Jeffrey 1983 and, applied to the present case, Grafstein 2002: 158–9. A propositional analysis is standard but Bradley (1998), among others, uses sentences. On another technical note, there is nothing contrived about using conditional as opposed to unconditional probabilities in formulating the three stochastic facts listed in the text. We follow Field (1977: 381), who follows Karl Popper, in treating conditional probabilities as primitives. This requires $\mathrm{Prob}(B|A) \geq 0$ for all propositions A and B (see Harper 1975: 232, Axiom a1). We can then define $\mathrm{Prob}(A) \equiv \mathrm{Prob}(A|T)$, where T is any certain proposition. In this formulation, the classical definition of the conditional probability $\mathrm{Prob}(B|A)$ can be derived only when $\mathrm{Prob}(A) > 0$. $\mathrm{Prob}(\mathrm{Cooperate}_1) > 0$ is a natural assumption here, since if $\mathrm{Prob}(\mathrm{Cooperate}_1) = 0$, the agent's decision has already been determined. Levi (1992) questions the coherence of unconditional probabilities of decisions.

$$\text{Prob}(\text{Defect}_2|\text{Defect}_1) > \text{Prob}(\text{Cooperate}_2|\text{Defect}_1), \tag{2}$$

$$\begin{aligned}
&\text{Prob}(\text{Cooperate}_2|\text{Cooperate}_1) \times U_1(\text{Cooperate}_1, \text{Cooperate}_2) + \\
&\quad \text{Prob}(\text{Defect}_2|\text{Cooperate}_1) \times U_1(\text{Cooperate}_1, \text{Defect}_2) \geq \\
&\quad \text{Prob}(\text{Cooperate}_2|\text{Defect}_1) \times U_1(\text{Defect}_1, \text{Cooperate}_2) + \\
&\quad \text{Prob}(\text{Defect}_2|\text{Defect}_1) \times U_1(\text{Defect}_1, \text{Defect}_2).
\end{aligned}$$
$$\tag{3}$$

Obviously, the situation is symmetric: players 1 and 2 can be interchanged. From this point on, we will assume, unless otherwise noted, that inequalities (1), (2), and (3) are true.

CEU theorists take inequality (3) to be a necessary and sufficient basis for cooperation. CDT per se does not dispute any of these inequalities, which are rooted in the beliefs of the players. But regarding inequality (3) in particular, CDT wants us to "hold that thought" on the grounds that this inequality does not provide an actionable guide to the best decision (that is, correlation does not imply causation). According to CDT, the stochastic dependencies at issue should not inform players' decisions because these facts do not control the behavior they describe. Therefore, CDT continues to recommend Defect as the dominant strategy.

In short, CDT does not seek to dispute the stochastic facts so much as defang them. The immediate problem with this attempt to maintain the consistency of the statistical facts (1), (2), and (3), on the one hand, and the pursuit of the dominant strategy, on the other, is that the latter undercuts the former. Insofar as all players are commonly known to be rational and are playing the dominant strategy, they render inequality (1) untrue since

$$\text{Prob}(\text{Cooperate}_2|\text{Cooperate}_1) = 0.^6$$

Similarly,

$$\text{Prob}(\text{Cooperate}_2|\text{Defect}_1) = 0.$$

Inequality (3) then reduces to

$$U_1(\text{Cooperate}_1, \text{Defect}_2) \geq U_1(\text{Defect}_1, \text{Defect}_2) \tag{3'}$$

which, of course, is false.

---

[6] In standard game theory, players not only behave rationally but know that all other players behave rationally, know that the other players know that they too play rationally, and so on (for the definition of common knowledge, see Fudenberg and Tirole 1991: 542–43); although compare Aumann and Brandenburger 1995: 1161–2.

It is worth interjecting that opponents of CDT have used the opposite example of degenerate probabilities – $\text{Prob}(\text{Cooperate}_2|\text{Cooperate}_1) = \text{Prob}(\text{Defect}_2|\text{Defect}_1) = 1$ – to test the limits of what CDT advocates will stomach (e.g., Levi 1975; Seidenfeld 1985). After all, when a player is absolutely certain that her cooperation is associated with higher utility than her defection (the modifier *expected* is superfluous), there is simply no longer any wiggle room for expecting any other result. Perhaps the case of $\text{Prob}(\text{Defect}_2|\text{Cooperate}_1) > 0$, even when it is very small, encourages players to believe they can beat the odds. But the hope of beating impossible odds is a tougher sell.[7]

In any case, while the appeal to degenerate probabilities may make life particularly difficult for CDT advocates, at the same time it seems to sacrifice all the advantages of focusing on PD instead of Newcomb's Problem. Both setups now seem totally contrived. The important lesson of inequality (3′), however, is that for CDT there is no logical space between the plausible stochastic assumptions (1) and (3) and the implausible assumption $\text{Prob}(\text{Cooperate}_2|\text{Cooperate}_1) = \text{Prob}(\text{Cooperate}_2|\text{Defect}_1) = 0$ when all players are rational as defined by CDT. CDT as a social scientific theory is incompatible with assumptions (1) and (3). In other words, CDT cannot rest content with the attempt to defang assumptions (1) and (3). It must deny them as possibilities.

It is worth emphasizing this critical point. According to Kanazawa (2000: 436), "Grafstein's voters therefore perceive an illusory correlation between their present behavior (voting versus abstention in the current election) and the simultaneous behavior of other voters (*their* voting or abstention)." But why must the correlation be illusory? Although not rooted in direct behavior-to-behavior causal dependencies, these statistical dependencies are, by assumption, true. If causal independence implied statistical independence, there would be no spurious correlations for statisticians to investigate. In short, if CDT is committed to denying assumptions (1) and (3), it is insisting on very strong empirical requirements, not merely shedding illusions.

In defense of these strong empirical requirements, one could argue, following Fischer (1999), that in terms of objective chances the probability of a voter's vote deciding the election is 0 or 1 since the remaining electorate is evenly divided or not. CEU theory, of course, focuses on subjective probabilities, foreswearing any claim to having a God's-eye view.

There are other ways to push back against the degenerate probabilities counterexample to CDT. Perfection, one might respond, is not of this world,

---

[7] Ahmed (2015) analyzes, critically, interpretations accepting a disjunction between $\text{Prob}(\cdot|\cdot) = 1$ cases and $0 < \text{Prob}(\cdot|\cdot) < 1$ cases.

and therefore perfect predictability is not a suitable working assumption for a realistic social scientific theory. Some extensions of standard game theory, for example, recognize the possibility of "trembles," small "mistakes" producing out-of-equilibrium behavior that otherwise would be theoretically impossible (e.g., Selten 1975). If trembles are assumed, $\text{Prob}(\text{Cooperate}_2|\text{Cooperate}_1) \neq 0$ ever; neither does $\text{Prob}(\text{Cooperate}_2|\text{Defect}_1)$.

Does this offer the relief valve CDT needs? Technically this assumption does block the reduction of assumptions (1), (2), and (3) to (3'). Of course, perfection is not of the world of science either: we also cannot absolutely know for certain that backward causation or action at a distance between players is impossible, which counts against the causal argument that choosing Cooperation is necessarily irrational. Regardless, unavoidable trembles do not free agents from their stochastic dependencies.[8]

Jacobi (1993) identifies an apparently crucial discontinuity when the decision-maker does in fact achieve complete certainty.[9] At this point, according to Jacobi, rational choice is rendered meaningless. The only way perfect predictability could be achieved is if (i) the players causally interact, and this interaction is deterministic, or (ii) a common cause deterministically controls the behavior of both players. In either case, according to Jacobi (1993: 12), "Rational choice is, by any pragmatic definition, completely absent from these situations, although [the decision-maker] may *subjectively* experience the act of choosing."

Once again, if probabilities are subjective à la CEU theory, players subjectively experience the key conditional probabilities associated with PD – $0 \leq \text{Prob}(\text{Cooperate}_2|\text{Cooperate}_1)$, $\text{Prob}(\text{Defect}_2|\text{Defect}_1) \leq 1$ – in the same sense they experience the act of choosing. They are not out of their minds when the inequalities are strict. If, on the other hand, these conditional probabilities reflect objective chances, they are part of the causal machinery of the world. If so, then presumably they are impervious to any supernatural freedom exercised by the players. Players must adapt to them.

The *artificial* contrast between degenerate and nondegenerate conditional probabilities aside, the distinction between epistemic and objective probabilities invoked by some CDT advocates (e.g., Gibbard and Harper 1978: 154–55) revives the Defect option in PD only if players can bootstrap

---

[8] An alternative approach is to think of seemingly irrational trembles as the product of rational choices. This approach makes it difficult even to define rationality and, overcoming that, to determine when individuals satisfy the definition (e.g., Binmore 1988, 1993).

[9] Jacobi focuses on Newcomb's Problem; the discussion in the text translates his analysis into the PD context. Ahmed (2014a: 170–79) addresses the "discontinuity" position at length.

themselves outside their own beliefs, only if they can will themselves not to believe what they believe. Put less dramatically, they must come to believe – truly believe – the causal account. The causal account insists that their behavior has no impact on the behavior of others in PD. Therefore, if all players are rational in the CDT sense and believe their fellow players are rational in this sense as well, then instead of inequalities (1) and (2) the players believe, among other things:

$$\mathrm{Prob}(\mathrm{Cooperate}_2|\mathrm{Cooperate}_1) = \mathrm{Prob}(\mathrm{Defect}_2|\mathrm{Cooperate}_1) \quad (1')$$

$$\mathrm{Prob}(\mathrm{Defect}_2|\mathrm{Defect}_1) = \mathrm{Prob}(\mathrm{Cooperate}_2|\mathrm{Defect}_1). \quad (2')$$

Obviously, inequalities (1) and (1') are incompatible, as are inequalities (2) and (2'). We are back to the conclusion that CDT is unable to abide by the original terms of the debate. Of course, (1') and (2') do not imply complete ignorance: it may be that player 1 believes that $\mathrm{Prob}(\mathrm{Defect}_2) \neq \mathrm{Prob}(\mathrm{Cooperate}_2)$ in addition to inequalities (1') and (2').

3 Normative Issues

I have asked the reader to assume for the sake of argument that decision-makers are in fact rational, however the reader defines rationality. At least as far back as Friedman (1953), many social science adherents of rational choice theory have ducked the realism issue by claiming they assume only that individuals behave *as if* they are rational. Strictly speaking, this could mean that rationality is simply a property of behavior. Behavior, in turn, would be understood as a vector of belief and desire, each of these understood as embodied in the decision-maker's brain. Rational choice theorists, however, usually mean by *as if* rationality that subjects are not assumed to be thinking consciously in the way rational choice theory models them. Rational decision-makers are still assumed to choose in some substantively important sense, although not necessarily by bringing together belief-desire components in the specific way standard rational choice models indicate.

With conscious rationality gone, why do rational choice theorists typically retain this last vestige of the original naïve, flat-footedly realist rational choice theory? Evidently, the idea is that without a residual notion of true choice, Jacobi's (1993) worst fears would be realized. In the same vein, Elster (1989: 6) writes, "One cannot be rational if one is the plaything of psychic processes that, unbeknownst to oneself, shape one's desires and values."

Elster (1989) would have us choose our own preferences. On what basis? Our preferences, with the resulting infinite regress? The uncaused causal self simply willing them? His hope is particularly problematic if, as Dennett (1984: 78) argues, "We do not witness [our decision] being *made*, we witness its *arrival*." Or compare Nietzsche (2001: Ch. I s. 17): "a thought comes when 'it' wishes, and not when 'I' wish."

True, Dennett is a compatibilist (see also Dennett 2003): he does not deny free will just because of scientific determinism. But this is not the kind of free will at issue here, "the ultimate, buck-stopping responsibility for what we do" in Strawson's (2003) nice formulation. Nor, Dennett agrees, does scientific indeterminism introduce any additional wiggle room for the free-willing self. If he is correct, the absence of buck-stopping responsibility haunts any decision-makers trying to evade the stochastic facts characterizing their behavior. It's like a nightmare: turn left, the stochastic demon is still on your trail; dodge right, still following you. There is no Archimedean point from which to survey these stochastic dependencies in order to decide whether or not to embody them, whether or not to shed their applicability.

Although our focus is on empirical, not normative theory, I want to pause to consider the normative policy implications of this loss of freedom. For without straying at least this much from our empirical mission, I risk encouraging the continuing suspicion that I have not fully addressed the residual normativity associated with rationality that I identified in section 1. In other words, rather than addressing Newcomb's Problem social scientifically, as advertised, one might conclude I simply changed the subject.

If decision-makers lack free will (again, in the buck-stopping sense), what possible advantage does policy advice offer policymakers or, for that matter, individuals facing PD? Policy advisers, of course, can provide new information to their advisees: "An adviser has to know something the advisee doesn't know" (O'Flaherty and Bhagwati 1997: 215). A fully developed theory of rationality encompassing information processing limitations would also encompass a theory of rational inattention (e.g., Sims 2003). A fully developed model of rational inattention, in turn, would incorporate policy advisers who fill in the policymaker's information gaps. This model, however, should not grant advisers any greater fundamental freedom than it should grant their advisees.

The previously addressed distinction between deterministic and stochastic setups has played a role in theoretical discussions of policy advice. The original context was the so-called policy-ineffectiveness proposition (Sargent and Wallace 1976). According to this proposition, policymakers cannot use monetary policy to move the economy by exploiting the Keynesian trade-off between inflation and employment. These attempts are thwarted by market actors who anticipate the consequences of any policy adopted and thereby negate its real economic impact.[10]

Market actors can bring about this standoff because they know the policymakers' money supply rule, including its parameter values. In this context, the policy adviser's information becomes part of the overall policymaking process and therefore part of the model market actors deploy. There is no privileged position for the would-be policymaker: "One cannot analyze the choice of policy variables without cutting through the seamless web of a model in which all policy variables are determined inside the model" (Sims 1986: 3).

Market actors can neutralize the impact of a policy *rule* they know or have inferred. This is not the same as restricting the impact of policymakers who introduce policy *innovations* (LeRoy 1995). In the space between settled monetary regimes, market actors no longer have available a data-tested rule to leverage. Perhaps policy advisers and policymakers can have an impact on the real economy during these transitional periods.

Sargent and Wallace's (1981: 213) rejoinder to this kind of suggestion is important: "In order for a model to have normative implications, it must contain some parameters whose values can be chosen by the policymaker. But if these can be chosen, rational agents will not view them as fixed and will make use of schemes for predicting their value... [T]hose parameters become endogenous variables." Therefore, "complete rationality seems to rule out ... freedom for the policymaker" (Sargent and Wallace 1981: 211).

When it comes to free choice, policymakers do not fare any better than those their policies target, who do not any fare better than decision-makers facing Newcomb's Problem. Decision-makers of all kinds are enmeshed in a network of stochastic dependencies they struggle in vain to escape.

Of course, hope springs eternal: "Even if we could read the book of our lives we would have to decide whether to believe it – whence Newcomb's Problem" (O'Flaherty and Bhagwati 1997: 213). To begin making sense of this odd decision, let us recognize that rationality is neither a necessary nor a sufficient condition to believe the book of our lives. The necessary part is obvious. As for sufficiency, we may find ourselves approaching the decision to believe the

---

[10] Since the money supply rule analyzed by Sargent and Wallace (1976) includes an additive disturbance term, market actors can only foil policymakers in expectation. The realization of these policies will exhibit random deviations from market expectations. It is worth noting that my discussion of the policy-ineffectiveness proposition is designed to make a theoretical point, not to make a claim about its empirical credibility.

book of our lives with very tight priors preventing a rational belief in the book. If it is the book of *our* lives, however, it not only tells us what decisions we made when facing Newcomb's Problem. The book describes the future decisions we will make, maybe after mulling them over and thrashing them out for a while, and however reluctantly.

Our decision about whether to believe the book does not change the book (since it's already at the printer, it's causally independent of our beliefs, one might say). Imagine the book of your life says that in the near future a truck will be headed toward you, but you will decide to jump out of the way. Will you decide instead to die just to spite the book of your life? No, you will act out the script. Anyway, however you wish to interpret a suicidal decision, it would not begin a new chapter in the book of your life. It would represent a futile attempt to start a new book that doesn't belong to you.

You can run but you can't hide:

> Endogenization of the sort that frustrates advice giving lurks closer to home. Economists and game theorists have themselves proposed models where agents are assumed to be rational not only by the theorists but by the agents themselves. Some stunning results are obtained on the assumption that players in games have common knowledge of rationality. I suggest that when assumptions like these are built into models of interaction between rational agents, we cannot finesse the problem for normative economists without dismantling the models (Levi 1997a: 224).

Levi would dismantle the models. I would dismantle the underlying assumptions about free choice that make the implications of these models so disconcerting. True, the common knowledge assumption is often unrealistic, and weakening the cognitive powers of agents would make these agents more susceptible to the unanticipated intrusions of their policymaking superiors. I suggest, however, that the mere theoretical possibility of a world in which the assumption holds tells us what we need to know about the intellectual viability of the free choice assumption.

It is time to confront an obvious question: if choice loses its role in decision-making – if this last vestige of the original, naïve theory of rationality is surrendered – what becomes of rational *choice* theory? Nothing in a technical sense happens. Obviously, our interpretation of the theory, or the individual models associated with it, changes. Under the CEU interpretation, when players are rational, their behavior as a matter of fact maximizes their conditional expected utility. We do not appeal to individuals who choose or strive to do this although they could have done otherwise. Up to any irreducible uncertainty built into our models of their behavior, they behave as modeled, as if they are maximizing CEU.

Our CEU interpretation could go deeper. We could characterize the processes by which rational decision-makers arrived at the beliefs guiding their current decisions and the processes by which these beliefs in turn were integrated with their preferences to determine a specific action. This interpretation would take the same stance toward these internal processes that we take toward external behavior: the processes represent a realization of whatever constraints rational choice theory imposes. No homunculus within oversees these internal processes, no separate self guides the external behavior.

Finally, it is worth confronting a legitimate concern about the feasibility of using CEU theory to apply game theory social scientifically: game theory's foundation seems to assume the independence of states of the world and acts that CEU theory brings into question. In appealing to CEU interpretations of game theory, do I want to have my cake and eat it too? It is worth noting in this regard that Aumann and Brandenburger (1995) have reconceptualized game theory to provide a better motivation for game theoretic equilibria and, in the process, have made the theory hospitable to CEU theory.

Specifically, Aumann and Brandenburger's (1995) proposed game theoretic architecture departs from the tradition of Savage (1972), who treats states of the world as causally independent of the decision-maker's acts.[11] Aumann and Brandenburger (1995: 1174) dispute this assumption's relevance for "the interactive, many-person world" that social scientists study. Put simply, acts are part of the state of the world: "Our world is shaped by what we do." Needless to say, causal dependence implies stochastic dependence, so we are in CEU's orbit as well. The upshot is that, in Aumann and Brandenburger's (1995) formulation of game theory, the stochastic dependence between a player's act and the state of the world the player inhabits is central to understanding the strategic interdependence characteristic of games: "The assumption of [stochastic] independence, which is normally accepted without discussion in non-cooperative game theory, in fact becomes quite questionable once the decision-theoretic viewpoint is adopted" (Brandenburger 1992: 89; see as well Aumann 1987: 16–17).[12]

---

[11] Specifically, his representation theorem for expected utility weights the utility of each outcome by the *unconditional* probability of the relevant state of the world. For a discussion of Savage's framework and its ramifications for Newcomb's Problem, see, e.g., Ahmed 2014a: 16–34.

[12] In effect, the agent's act represents a choice among the different states of the world incorporating the agent's alternatives. In the formal development of their approach, however, Aumann and

4 More Social Science Applications

As I discussed in section 1, CEU theory gains a potential explanatory advantage over CDT when there are stochastic dependencies among players' behaviors, that is, stochastic dependencies remaining after conditioning on any other relevant factors. The differences between the theories, nevertheless, can still amount to a distinction without a difference. These dependencies must be sufficiently large and the utility differences among alternatives must be appropriately aligned. This makes CEU theory an empirical theory.

Of course, the sufficiency of a CEU explanation of important and empirically challenging behavior like turnout does not make it necessary. There are other possible explanations for this and related behavior, each with its own strengths and weaknesses. Yet because CEU explanations are constructed at a fundamental micro level and, I have argued, are consistent with the rationality assumption, they have a potential breadth, flexibility, and internal consistency that justify their receiving the special consideration we have accorded them here.

Let us return, then, to the example of turnout in mass elections discussed briefly in section 1. The importance of the social scientific challenge turnout poses when participation is not compulsory is due not only to the importance of democratic elections, which are certainly important enough. Thinking more broadly, exercising the franchise is representative of the many ways in which people do not behave as isolates or as solipsistic automata. People seem to behave in awareness of their connections to others, not necessarily the whole human race, as moralists would prefer, but nonetheless in communitarian or tribal ways. They seem to draw part of their identity from their membership in groups.

Another electoral example will illustrate how CEU theory grounds a more nuanced analysis than the simple, qualitative example of turnout in mass elections might suggest. Consider situations in which voters might be tempted to vote strategically: US multicandidate primary elections or multiparty elections elsewhere. In these situations, some voters may conclude that their

Brandenburger (1995) fail to appreciate that the new states of the world the agent triggers encompass these new acts and all the stochastic dependencies associated with them. Specifically, Aumann and Brandenburger do *not* define a rational act as one having a higher expected utility in the state of the world incorporating that act than any alternative state of the world encompassing an alternative act that is feasible for that agent. Rather, the rational act has a higher expected utility than any other act given the state of the world associated with the *original* act. Somehow, the essential role of acts as part of the definition of states of the world is sacrificed. Grafstein (2002: 158–63) formally explores these points.

preferred candidate or party stands such a small chance of winning that they choose to vote for a less preferred alternative.

Abramson et al. (1992) explore the extent to which a candidate's viability – likelihood of winning – affects strategic voting in US primaries. Their idea is that "the rational voter's choice depends on the comparative *utilities* associated with the candidates and the relative *probabilities* of outcomes" (1992: 56). "The heart of expected utility is the multiplication of utility and probability; and we construct such models here by multiplying the difference in viability of two candidates by the difference in utility evaluations" (1992: 62). Analyzing survey data from the 1988 US presidential primaries, they estimate the impact on the voter's vote choice using the explanatory variable

$$PB_{jk} \equiv \left(P_j - P_k\right)\left(U_j - U_k\right), \tag{4}$$

where P_j (respectively P_k) represents the survey respondent's assessment of candidate j's (respectively k's) probability of winning the primary; U_j (respectively U_k) represents the utility associated with j (respectively k) as measured by the candidate's score on a 100-point feeling thermometer; and $j < k$ implies that $U_j > U_k$. Accordingly, Abramson et al. (1992: 62) "expect … estimates for PB_{12} and PB_{13} to be positive, since greater expected utility for the most preferred candidate compared to the other two should make supporting the most preferred candidate more likely."

As Abramson et al. (1992: 62) note about their empirical findings, the signs and magnitudes associated with the key explanatory variables for their respondents' most preferred candidates – PB_{12} and PB_{13} – are precisely those suggested by their interpretation of expected utility calculations related to strategic voting. The problem is that the findings should not have been clear and decisive since equation (4) provides an incorrectly formulated operationalization of CDT's expected utility theory (see Grafstein 2003).

P_j and P_k are the source of the problem. Recall that, in keeping with expected utility theory, these represent the *unconditional* probabilities of winning that respondents assign to candidates j and k respectively. But CDT voters care about the probability of being decisive by making or breaking what otherwise would be a tie vote. Therefore, only for $(P_j - P_k) < 0$ is it true that the voter's expected utility from casting the appropriate vote is increasing in $(P_j - P_k)$. For $(P_j - P_k) > 0$, the voter's expected utility of voting is decreasing in $(P_j - P_k)$, unless P_j is so much smaller than the third-ranked candidate's that the vote is irrelevant. The inequality in the first case is consistent with Abramson et al.'s (1992) empirical findings; the inequality in the second case is not. For their findings to be consistent with expected utility theory when the two cases point in opposite directions, the signs and magnitudes of their results should have been weak and ambiguous.[13]

CEU theory, I believe, offers a better explanation of the Abramson et al. (1992) results. The survey respondents Abramson et al. (1992) study offer not only estimates of the candidates' probabilities of victory and the utility readings U_j and U_k. They also declare a vote intention. Though elicited separately, presumably all these elements of the respondent's evaluation are held simultaneously. If so, the respondent's probability estimates are made against the background of the respondent's vote. Therefore, I propose interpreting P_j and P_k as the subjective *conditional* probabilities of victories by j and k respectively, given the respondent's actual vote (for j or k; the third-ranked candidate is a dominated choice). A little manipulation of equation (4) yields

$$PB_{jk} = P_j(U_j - U_k) + P_k(U_k - U_j).\qquad(5)$$

With the CEU interpretation in mind, equation (5) characterizes the CEU of voting for the candidate the respondent did choose, with P_j and P_k appropriately conditioned on the respondent's actual vote and utilities calculated net of opportunity costs. Now, whether $P_j > P_k$ or $P_j < P_k$, PB_{jk} *is* an increasing function of $P_j - P_k$, just as the empirical findings suggest. A model that was empirically well supported is restored to internal coherence.

To be clear, equation (5) highlights the CEU interpretation of the equation Abramson et al. (1992) actually estimated, equation (4). There is no numerical difference between the values of the variables P_j and P_k in the two equations. Only the interpretations – unconditional versus conditional – differ. But under the latter interpretation, for a j voter, the larger P_j is relative to P_k, the more appropriate that vote is; similarly, for a k voter, the larger P_j is, the less appropriate that vote is.[14]

---

[13] In fairness, the American National Election Studies, the source for their data, does not ask respondents to estimate the probability their votes are decisive; Abramson et al. (1992) use the only estimates available. In his study of strategic voting, Cain (1978: 647) addresses this important inconsistency between the inequalities by adopting a counterpart formula for equation (4) using the transformation $(P_j - P_k)^{-1}$. In good CDT fashion, Cain (1978: 641) writes that these probabilities are calculated based on "the likelihood that the vote will be decisive (i.e., affect the outcome)." Nevertheless, in good CEU fashion, he also characterizes these probabilities as "the perceived conditional probabilities that an individual's preferences will be elected given that this individual votes sincerely."

[14] Note that P_j and P_k are inversely related since the Abramson et al. (1992) regression using PB_{jk} holds constant the probability estimate for the third-ranked candidate. Incidentally, if we had the luxury of doing the estimations from scratch, it would be better not to invoke opportunity costs.

In this example, we have used CEU theory diagnostically to explain strategic voting behavior. We have not confronted the lurking empirical question whether these are rational voters we are studying. Obviously, the utility of CEU theory and the nature of rationality are distinct issues. The theory, after all, could help us explain how people behave even if they are irrational. Or put more pointedly, CEU theory could work precisely because people are not particularly rational.

In short, this chapter could have focused on applications of CEU theory inspired by the ways Newcomb's Problem can be translated into social scientific problems like PD, sincere voting in mass elections, and strategic voting in multicandidate elections; and it simply could have ignored the rationality issue altogether. This pragmatic approach would show greater respect for the division of labor between social science and philosophy. Without getting too far afield, I want to offer two reasons – not fully developed arguments – for violating this disciplinary boundary, one social scientific and one philosophical.

Why is it useful for social scientists to assume the people they study are rational? Empirical deviations from any meaningful standard of rationality certainly occur. Perhaps they are inevitable. Agents who behave rationally, however, do not make systematic mistakes. Therefore, their behavior and the patterns created by the aggregate behavior they generate are not inherently fragile, sustainable only in the absence of self-reflection, reconsideration, or the awareness of existing opportunities to make the agent better off. When agents are rational, these fundamental objects of social scientific study are stable or in equilibrium. This general point was implicit in our earlier discussion of the role of policy advice when agents are rational.

Extending the equilibrium concept to a more natural setting, we would prefer to have models of social interaction that are internally consistent in the sense that the behavior of individuals populating a given model does not disturb the pattern being modeled, and the pattern being modeled does not force agents in the model to change their behavior. There certainly are equilibrium models with irrational or semi-rational agents, particularly regarding their expectations and the way they learn to update their beliefs.[15] So, in no way am I asserting the necessity of rational choice assumptions for equilibrium models. Nevertheless, when model agents are rational, the

---

[15] Thomas Sargent, a pioneer of rational expectations analysis, has also pioneered the analysis of models in which agents do not have rational expectations, but must learn about their changing and emerging environment (e.g., Sargent 1993).

equilibrium the model identifies is not subject to revision and the model subject to failure whenever these agents better understand the pattern of equilibrium behavior they encounter or more effectively process information they receive. The equilibrium enjoys a welcome degree of stability.

Ultimately, the impetus for rational choice models in social science, beyond practical empirical results, is the advantage of working with agents who have a clear connection to reality. Reality is not systematically fooling or confusing them. Philosophers have noted the same advantage of employing models of human psychology in which behavior is connected to beliefs and desires in a transparent way (e.g., Broome 1990b, 1991a, 1991b). Thus, given a systematic conflict between stated beliefs or preferences and actual behavior, we are inclined, at some point, to declare that "actions speak louder than words" and refuse to let the words speak for themselves. In light of the agent's actions, we embrace rational consistency by revising our interpretation of these stated beliefs or preferences. In the extreme case, when faced with subjects who exhibit rapidly cycling patterns of behavior – preferring A to B to C to A – or, worse, rapid preference reversals – preferring A to B to A – we are inclined to declare that such people are not so much unable to make up their minds as lack minds to make up.

According to this argument, our folk belief-desire psychology is structured in a rationalistic way. The rationality assumption makes behavior constitutive of what we take an agent's preferences and beliefs to be (e.g., Dennett 1987; Davidson 2001). If, then, we surrender the assumption of rationality as an unnecessary "fiction," we run the risk of throwing out the folk-psychology baby with the behavioral anomaly bath water. For we cannot even casually theorize about preferences and beliefs when these psychological elements become too detached from the rational choice framework: "To see too much unreason on the part of others is simply to undermine our ability to understand what it is they are so unreasonable about" (Davidson 2001: 153).

5 Concluding Remarks

In Newcomb's Problem an external agent predicts the decision-maker's behavior with sufficient accuracy that the CEU of choosing one box exceeds the CEU of choosing two. In PD, a parallel and interesting case arises in which the CEU of choosing Cooperate exceeds the CEU of choosing Defect. In both cases, the decision-maker's expected utility is conditional on the decision-maker's own behavior, among other things. In both cases, the rational decision-maker should follow his or her expectations.

Using CEU theory to connect Newcomb's Problem and PD opens up a world of empirical possibilities and philosophical controversies. One plausible way to grapple with these challenges is to segregate the social scientific and philosophical analyses of both. Yet while the appropriate social scientific analysis of PD does not dictate the appropriate philosophical analysis of Newcomb's Problem, and likewise for the converse, analyzing decision-makers as CEU maximizers, I have argued, is a productive empirical research project; and what makes it promising, in part, is that CEU maximization is the core of a plausible account of rationality. No guarantees, but the basis for a continuing dialog.

6   Success-First Decision Theories

Preston Greene

1 Introduction

The classic formulation of Newcomb's Problem is often thought to compare evidential and causal conceptions of expected utility, with those maximizing evidential expected utility tending to end up far richer than those maximizing causal expected utility. Thus, in a world in which agents face classic Newcomb Problems, the evidential decision theorist might ask the causal decision theorist: "If you're so smart, why ain'cha rich?" Ultimately, however, the expected riches of evidential decision theorists do not vindicate their theory, because their success does not generalize. For example, in a world in which agents face "transparent" variants of Newcomb's Problem, where the contents of both boxes are revealed before a choice must be made, both causal and evidential decision theorists tend to end up poor.[1] Thus, in such a world, those following some other theory could ask of both causal and evidential decision theorists: "why ain'cha rich?"

Consider a theory that allows the agents who employ it to end up rich in worlds containing both classic and transparent Newcomb Problems. This type of theory is motivated by the desire to draw a tighter connection between rationality and success, rather than to support any particular account of expected utility. We might refer to this type of theory as a "success-first" decision theory. The main aim of this chapter is to provide a comprehensive metatheoretical justification of success-first decision theories as accounts of rational decision-making.

The desire to create a closer connection between rationality and success than that offered by standard decision theory has inspired several success-first decision theories over the past three decades, including those of Gauthier

---

[1] The transparent Newcomb Problem is discussed in Gibbard and Harper 1981: 181–2. For further examples of a why-ain'cha-rich objection lodged against EDT, see Arntzenius 2008: 289–90 and Soares and Fallenstein 2015: 2–3. See Lewis 1981b and Joyce, 1999: 151–4 for other types of responses to the why-ain'cha-rich objection on behalf of CDT.

(1986), McClennen (1990), and Meacham (2010), as well as an influential account of the rationality of intention formation and retention in the work of Bratman (1999). McClennen (1990: 118) writes: "This is a brief for rationality as a positive capacity, not a liability—as it must be on the standard account." Meacham (2010: 56) offers the plausible principle, "If we expect the agents who employ one decision making theory to generally be richer than the agents who employ some other decision making theory, this seems to be a prima facie reason to favor the first theory over the second." And Gauthier (1986: 182–3) proposes that "a [decision-making] disposition is rational if and only if an actor holding it can expect his choices to yield no less utility than the choices he would make were he to hold any alternative disposition." In slogan form, Gauthier (1986: 187) calls the idea "utility-maximization at the level of dispositions," Meacham (2010: 68–9) a "cohesive" decision theory, McClennen (1990: 6–13) a form of "pragmatism," and Bratman (1999: 66) a "broadly consequentialist justification" of rational norms.

Even though these metatheoretical views have some initial plausibility, they have not been rigorously developed past the slogans (the object-level theories, which I evaluate in section 3.3, are much more developed), and it is partly for this reason that many theorists working on practical rationality view success-first decision theories with deep suspicion. In section 2, I provide a comprehensive metatheory for success-first decision theories, which I locate inside an *experimental approach* to decision theory. The experimental approach helps reveal what the why-ain'cha-rich objection is really getting at, and it provides a guiding light for success-first decision theory. I show that the approach is compatible with the general ambitions of both CDT and EDT, although here my primary focus is on the causalist. The experimental approach diverges from the standard methodology of CDT by suggesting that experiments, rather than mathematical axiom systems, are the proper tool for studying causal efficacy.

In section 3, I outline the sort of object-level theories that follow from the experimental approach. For the causal experimentalist, the difference between one- and two-boxing in Newcomb's Problem is not that between maximizing causal and evidential expected utility, but rather that between *acts* and *decision theories* serving as the *independent variable* of the experiment.

2 The Experimental Approach to Decision Theory

Decision theory is concerned with instrumental rationality: the rationality of instrumental aims. As such, it is not concerned with the rationality of ultimate aims. Many decision theorists point out that decision theory is relevant to all forms of decision-making: not just in prudential contexts, but in moral contexts as well, *precisely because* decision theory is not concerned with ultimate aims.

Accordingly, the decision theorist's job is like that of an engineer in inventing decision theories, and like that of a scientist in testing their efficacy. A decision theorist attempts to discover decision theories (or decision "rules," "algorithms," or "processes") and determine their efficacy, under certain idealizing conditions, in bringing about what is of ultimate value.

Someone who holds this view might be called a *methodological hypernaturalist*, who recommends an *experimental approach to decision theory*.[2] On this view, the decision theorist is a scientist of a special sort, but their goal should be broadly continuous with that of scientific research. The goal of determining efficacy in bringing about value, for example, is like that of a pharmaceutical scientist attempting to discover the efficacy of medications in treating disease.

For game theory, Thomas Schelling (1960) was a proponent of this view. The experimental approach is similar to what Schelling meant when he called for "a reorientation of game theory" in Part 2 of *A Strategy of Conflict*. Schelling argues that a tendency to focus on first principles, rather than upshots, makes game-theoretic theorizing shockingly blind to rational strategies in coordination problems. A simple example asks respondents to name "heads" or "tails" with the understanding that if a partner in another room does the same they both receive a prize (Schelling, 1960: 56). Since the heads-heads and tails-tails outcomes are both in equilibrium and Pareto-optimal, the game-theoretic answer is that agents should pick a response at random. This strategy secures the prize half of the time. But respondents tend to do much better than that. A large majority of respondents choose heads (Mehta et al. 1994). Schelling explains that "heads" represents a point of convergence of expectations for most people, which he calls a "focal point."[3] He introduces many examples to show that attention to focal points is a valuable tool for coordinating, bargaining, and deterring. Focal points, however, have received little attention from game theorists.[4]

---

[2] "Hyper" because many decision theorists think of themselves as methodological naturalists, but do not subscribe to the experimental approach, often due to concerns that I allay in section 2.2.

[3] In another famous example, Schelling asked respondents to pick a time and place to meet a partner in New York City without prior communication. Most respondents selected Grand Central Station at noon (Schelling 1960: 55–6).

[4] Exceptions include Sugden 1995 and Sugden and Zamarrón 2006.

Focal points would not disappear under idealization, and so cannot be dismissed as concerning non-ideal theory. Rather, the lack of attention to focal points is due to the difficulty in deriving a systematic theory of their use from axioms of rational choice. In contrast, Schelling imposes no such restriction on how a theory is derived. Instead, he commits to finding the best theory by *induction from its success*. Referring to coordination games like "heads/tails," Schelling (1960: 283–4, emphasis added) writes that his basic premise is that rational players realize "that some rule must be used if success is to exceed coincidence, and that the best rule to be found, *whatever its rationalization*, is consequently a rational rule."

The experimental approach to decision theory departs from standard methodology in the way imagined by Schelling. Rather than attempting to justify theories through deduction from a priori axioms, the main justification lies in the results. Only after expected results are determined should foundations be inferred. To borrow from William James (1912: 726): the strength of the standard decision theorist's system "lies in the principles, the origin, the *terminus a quo* of his thought; for us the strength is in the outcome, the upshot, the *terminus ad quem*. Not where it comes from but what it leads to is to decide."[5]

2.1 Experiments

The best metatheory for success-first decision theories develops Schelling's insights into a rigorous experimental approach. First, note how experiments in decision theory differ from those of other scientists. The decision theorist uses thought experiments, not concrete experiments, and they determine the efficacy of decisions or decision theories from the specification of problems, rather than through controlled trials.

These methodological differences are due to decision theory's goal of instrumental value-maximization across possible worlds. In comparison, the goal of treating a disease as it actually exists, for example, is severely circumscribed. Nevertheless, in each instance the function of thought experiments and concrete experiments is the same. In empirical science, experiments and thought experiments share a functional description (El Skaf and Imbert 2013).

An experiment aims to investigate how a *dependent variable* depends on an *independent variable*. For example, a research pharmacist attempts to discover medications (independent variable) and determine their efficacy in treating

---

[5] Quoted in Sugden and Zamarrón 2006: 620.

disease (dependent variable). This requires careful use of controls that help isolate the effects of the independent variable from others. It is not the role of the pharmacist to define "disease" and "cure," or even "therapeutic effect" and "adverse effect." Rather, the pharmacist is in the first place simply interested in determining the effects of a given treatment.

The same should hold of decision theory. As decision theory concerns instrumental rationality, it is not the place of the decision theorist to determine which states of the world are better than others. Rather, the decision theorist studies the efficacy of an action or decision theory (independent variable) in bringing about states of affairs (dependent variable). The value of these states is often taken as determined by the agent's preferences; however, in line with the idea that decision theory concerns all forms of decision-making, values can be determined in many ways. The idealizations (in idealized decision theory) act to isolate the effects of the independent variable from others, like an agent's belief accuracy or cognitive abilities.

Consider James Joyce's (1999: 146–7) fascinating description of New-comb's Problem:

> Suppose there is a brilliant (and very rich) psychologist who knows you so well that he can predict your choices with a high degree of accuracy. One Monday as you are on the way to the bank he stops you, holds out a thousand-dollar bill, and says: "You may take this if you like, but I must warn you that there is a catch. This past Friday I made a prediction about what your decision would be. I deposited $1,000,000 into your bank account on that day if I thought you would refuse my offer, but I deposited nothing if I thought you would accept."[6]

In Nozick's (1969) classic formulation, the bank account is an opaque box, and the money is in a transparent box. Joyce's otherwise equivalent formulation does an excellent job of highlighting the causal inefficacy of refusing the thousand dollars.

When determining causal efficacy under the experimental approach, there is a choice regarding the independent variable, and this is best highlighted by Newcomb's Problem. As Joyce's formulation reveals, if we set the independent variable to *acts*, then the causally efficacious option is to accept the thousand. We deduce, using causal dominance, that the effect of accepting (i.e.,

---

[6] Adapted from a version presented by Sobel (1985b: 198–9 n. 6). Sobel points out that the classic formulation of Newcomb's Problem can lead to confusions that make one-boxing seem more appealing.

"two-boxing") is to leave you a thousand richer than you would be had you chosen the other option. The good thing about causal dominance arguments is that they leave no room for doubt. If we set up this experiment and controlled for variables that are causally independent of acts, we would observe that everyone who accepts the thousand gets a thousand more than they would have had they chosen the other option. This all follows from the specification of the case.[7]

The story about causal efficacy remains the same in the transparent Newcomb Problem.[8] Suppose the psychologist informs his potential beneficiaries whether the money has been deposited before they decide, and as before, that he is an accurate predictor. Again, causal dominance applies. There are two possible conditions: you learn that the money has been deposited or it has not. In both conditions, you receive a thousand more by accepting than you would by refusing. This is what we would observe were we to set up the experiment and control for irrelevant factors. Thus, when studying causal act efficacy with the experimental approach, two-boxing is most efficacious in classic and transparent Newcomb Problems.

An experimental rationale can also be given for the split recommendations of evidential decision theorists who favor one-boxing in the classic Newcomb Problem but two-boxing in the transparent version. Rather than determining an act's causal efficacy, experiments can determine an act's 'foreseeable actual expected return' (Ahmed 2014a: 182 n. 34). To determine that, the relevant experiment would observe the average return of one- and two-boxing while controlling only for variables that are *evidentially* independent of the act (most importantly, the prediction is left uncontrolled). Since the predictor is accurate, these experiments would show a greater average return for one-boxing. Such an experiment would be "nonstandard" in that it would aim to establish correlation, rather than a causal relationship. However, this is unlikely to trouble evidential decision theorists, who

---

[7] Dominance arguments only apply if the states are independent of the available acts. Causal decision theorists hold that this is satisfied when the states are *causally* independent of the acts, as in Newcomb's Problem. Evidential decision theorists, however, require *evidential* independence of the acts, and, therefore, reject dominance reasoning in Newcomb's Problem (see Joyce 1999: 150–1). See footnote 10 for a type of dominance argument that is compatible with success-first decision theories.

[8] Gibbard and Harper (1978: 181) write: "Consider a variant on Newcomb's story: the subject of the experiment is to take the contents of the opaque box first and learn what it is; he then may choose either to take the thousand dollars in the second box or not to take it." Note that in their discussions of the transparent Newcomb Problem, Broome (2001: 101), Gauthier (1986: 157–89), and Parfit (1984: 7) all use "transparency" to denote the fact that the agent can be predicted, and not the fact that the agent can see into the boxes.

are already committed to the idea that rational decisions are not determined by causal relationships.[9]

Two possibilities remain: a recommendation to one-box in the transparent version but two-box in the classic version, and a recommendation to one-box in both. I know of no experimental rationale for the former, but the key, I believe, to understanding the motivation behind the latter is to recognize the impulse to study the causal efficacy of *decision theories* rather than acts.[10] Understood this way, the experiment should not test the effects of the act of one-boxing, as when studying act-efficacy, but rather of employing a decision-making theory that recommends one-boxing. Thus, we set the independent variable to decision theories, not acts, and imagine experiments that isolate the causal consequences of decision theories. That changes the story about who tends to end up with more money. As is assumed in the literature, those who employ a decision-making theory that recommends one-boxing tend to get a million, while those who employ a theory that recommends two-boxing tend to get a thousand. This follows from the specification of the case – what it means for a predictor to be "accurate," etc. Furthermore, these effects can be observed, in principle, in the same way that the effects of the different acts can be observed: by controlling for causally independent factors, applying accurate predictions to each type of agent, and noting the differences in wealth that result.[11]

Thus, from the causal experimentalist perspective, disagreement over New-comb's Problem is best attributed to disagreement over the appropriate independent variable, and not over the appropriate conception of expected utility. In fact, from that perspective, casting Newcomb's Problem as a

---

[9] What can be said against the evidential experimental approach? From my perspective, the largest hurdles are (i) the intuition that rational decision-making is a subjunctive endeavor in which the relevant considerations concern causal properties (Joyce 1999: 252–3), and (ii) the concern that this will result in a "one-box" recommendation for medical Newcomb Problems (see section 2.3.1). Whether (ii) is a problem depends on whether there exists a sound version of the "tickle defense" argument (see Ahmed 2014a: 91–7 for a recent example and further references).

[10] A type of dominance argument that is compatible with one-boxing in both versions is the "principle of full information" endorsed by Fleurbaey and Voorhoeve (2013: 114): "When one lacks information, but can infer that there is a particular alternative one would invariably regard as best if one had full information, then one should choose this alternative." This principle, when combined with the requirement of causal act-state independence, would merely require that a theory give the same recommendation in both classic and transparent Newcomb Problems.

[11] If one does not understand this motivation, and views act-efficacy as the only possible option for evaluating the rationality of decision-making, then one is likely to be befuddled by the many seemingly intelligent people working in fields related to artificial intelligence who view one-boxing in the transparent Newcomb Problem as rational. Arntzenius (2008: 290), for example, calls these people "insane." (The same applies, mutatis mutandis, to those befuddled by the many seemingly intelligent philosophers who view two-boxing in the classic Newcomb Problem as rational.)

disagreement over expected utility is a needless distraction. Endorsement of one-boxing is compatible with the general motivations of CDT and the use of causal notions to define expected utilities.

2.2 The Epistemic Problem with the Experimental Approach

The above might strike many as a nice idea, but unworkable as we move to more complex cases involving epistemic uncertainty. The problem is this. Most decision theorists believe that decision situations are defined by an agent's epistemic perspective, not the agent's actual situation. (This is the analog, in decision theory, of the idea that practical rationality concerns a "subjective ought," according to which what an agent ought to do depends on what the agent believes, not what the world is actually like). However, the experimental approach may seem compatible only with a more "objective" conception of decision situations,[12] since it is unclear how to experiment with an epistemic perspective. So, the experimental approach might seem hopeless as an account of subjective rationality.

This is the largest obstacle to the experimental approach, and one main reason knowledgeable theorists remain skeptical of success-first decision theories. Thus, the justification for the experimental approach or success-first decision theories *will never be complete* without a proper specification of the connection between an agent's epistemic perspective and the conditions by which the agent succeeds or fails.

The solution lies in an "experimental resolution" of the epistemic perspective. To create an experimental resolution, take a description of the agent's epistemic perspective, including probabilities, causal dependency hypotheses, etc.; in short, everything that is relevant to the decision situation. The *idealized experimental resolution* of the case is a specification of a situation that eliminates the effects of inaccuracies in the agent's epistemic perspective. Using this, we deduce the effects of the independent variable on average success and failure.

Some theorists will, of course, disagree about which epistemic factors are relevant to the decision situation. The experimental approach is neutral between these different conceptions. It only suggests that, for any proposed decision situation, the description be made detailed enough (if possible) to allow for experimental resolution.

---

[12] See Mellor 1991 for a defense of "objective decision theory."

In step form:

1. The case should describe in as much detail as necessary the agent's epistemic perspective.
2. Form an idealized experimental resolution of the case, controlling for the effects of inaccuracies or limitations in the agent's epistemic perspective.
3. Deduce the averages of good and bad states for different decisions or decision-makers under the experimental resolution.

This process is not as revisionary as it may seem. Theorists have long been forming idealized experimental resolutions of decision problems and making claims about them. Consider the orthodox claim that one-boxers tend to be richer than two-boxers. In descriptions of Newcomb's Problem, the predictor is often called "accurate," "reliable," or similar. If these terms are meant to describe the agent's epistemic perspective and not the actual situation, then the inference from the description to one-boxers' success is obviously invalid. Here, Nozick's (1969: 207, emphasis added) original presentation of the case is particularly apt. He writes: "Suppose a being in whose power to predict your choices you have enormous confidence" and "One might tell a longer story, but all this *leads you to believe* that almost certainly this being's prediction about your choice in the situation to be discussed will be correct." Notice that we cannot directly infer any actual average monetary payouts from this description.

When theorists describe the average success of one- and two-boxers, therefore, these claims *must* concern an idealized experimental resolution of Nozick's description, which controls for the effects of belief inaccuracy. Here is one such. To call the predictor reliable is exactly to say that there is a high chance of its predicting one/two-boxing in situations where one/two-boxing in fact occurs.[13] Under this assumption, the idealized experimental resolution demands a high chance of a one-box prediction given that one-boxing occurs, and a high chance of a two-box prediction given that two-boxing occurs. So, in the resolution, one-boxers tend to get a million dollars while two-boxers tend to get only a thousand (two-boxers do not *always* end up poorer, of course, and that is why the focus is on average payout across repeated hypothetical experiments, as in Step 3 above). The nature of the chances need not be metaphysically loaded: it could simply be a high hypothetical relative frequency of correct predictions. Something like this is necessary to make sense of the orthodox claim that one-boxers tend to be richer than two-boxers.

---

[13] Cf. Joyce 1999: 170 n. 36.

Can we always form experimental resolutions of well-formed decision problems? My main reply is that, ultimately, it does not matter. But let me first introduce the reasons for optimism about forming experimental resolutions.

Causal decision theorists share a common idea, and differ mostly on emphasis (Lewis 1981a: 5). Some, for example, emphasize chances in their formulations, and some counterfactual conditionals. But these are interrelated and, on some views, interdefinable. The idea that causal decision theorists share is that the agent's epistemic perspective yields a causal story that must be respected when determining rational decisions. This story is often called a "dependency hypothesis." Dependency hypotheses are formulated in various ways, but all involve the possibility of constructing an experimental resolution from the story.

For example, the experimental approach is particularly friendly to "chance" readings of dependency hypotheses as maximally detailed specifications of conditional chances.[14] For *objectivists*, these comprise propositions that do not vary from person to person, while for *subjectivists*, they are based on an agent's estimation of chances. Either way, we can create an idealized experimental resolution of the agent's decision situation by assuming that the conditions determining the agent's actual payout match the conditional chances under the agent's dependency hypothesis.

We can also formulate experimental resolutions using the counterfactual dependence version. At first glance, creating a usable experimental resolution from counterfactual conditionals may seem a challenge, since the relevant chances, which are required to calculate the averages of the states obtaining, are not supplied directly by the agent's epistemic perspective, but must be inferred from credences about counterfactuals. These inferences can seem especially difficult if the agent's credences are divided over conflicting counterfactuals. Nevertheless, epistemologists are doing promising research on the connection between credence and objective chance. Hájek (manuscript) suggests that credence "aligns with the truth" to the extent that it matches objective chances, and Mellor (1991: 274) proposes something similar. Other theories focus on when a credence is justified. Following Hájek and Mellor, perhaps a justified credence is one that matches or nearly matches objective chance, or one that is produced by a process that produces a high proportion of credences nearly matching objective chances. Following van Fraassen (1983, 1984) and Lange (1999), perhaps a justified credence is one

---

[14] See Joyce 1999: 166–7 for discussion.

that is the output of a process that is calibrated; i.e., a process that produces credences that match actual or potential frequencies. Or perhaps, as Tang (2016) suggests, expanding on Alston (1988, 2005), a justified credence is one that is based on some ground, where the objective probability of the credence having true content given that it is based on that ground matches the credence.[15] In each instance, we have a general formula for creating experimental resolutions: stipulate that the agent's credences are justified or at least align with the truth, and so match objective chances.

Even if one does not accept these theories of the justification of credences – for example, one prefers the concept of *accuracy*[16] – they can generate idealized experimental resolutions. The important point is that an agent's credences about propositions can be used in an idealized experimental resolution by stipulating that they match objective chances.

Now we come to the main point. The inability to construct idealized experimental resolutions of certain decision problems should not affect our evaluation of problems where it is possible to construct such a resolution. This is analogous to a common observation regarding decision-making under risk (where probabilities are known) versus decision-making under ignorance (where they are not). Even if one accepts a principle for decision-making under ignorance, such as minimax regret, one should not insist on it when probabilities are known. (Rather, in those cases, one should maximize expected utility.) Similarly for experimental resolutions: when an experimental resolution *can* be formed, there is no need to apply principles meant to deal with cases in which it cannot.

Furthermore, just as some believe that supposed decisions under ignorance can always be transformed into decisions under risk – perhaps via a principle of indifference – some may believe that one can always create experimental resolutions from well-formed decision problems. Others may believe that it is not always possible. An experimentalist about decision theory need not take a stand on this.[17]

---

[15] Beebee and Papineau (1997) make a similar proposal.

[16] See Joyce 1998. Also see Tang 2016: 74–6 for objections to this view.

[17] Examples of decision problems where it is difficult to construct idealized experimental resolutions suitable for studying act-efficacy involve *causal unratifiability*. For example, in Gibbard and Harper's (1978: 185–6) *Death in Damascus*, rational deliberation results in shifting calculations of the causal consequences of each act, and thus shifting idealized experimental resolutions for the decision problem. The same is true in "asymmetric" variants of *Death in Damascus*, in which one location is more pleasant than the other. (See also Egan's (2007) *Psychopath Button*. Interestingly, the suggestion that such cases lack an idealized experimental resolution suitable for

2.3 Further applications

We have seen how the experimental approach applies to the classic and transparent versions of Newcomb's Problem. Let us now apply it to the medical Newcomb Problem and to elucidating the difference between idealized and non-idealized decision theory.

2.3.1 Medical Newcomb Problems

Here is Andy Egan's (2007: 94) version of the medical Newcomb Problem:

The Smoking Lesion

Susan is debating whether or not to smoke. She believes that smoking is strongly correlated with lung cancer, but only because there is a common cause—a condition that tends to cause both smoking and cancer. Once we fix the presence or absence of this condition, there is no additional correlation between smoking and cancer. Susan prefers smoking without cancer

---

studying act-efficacy supports Joyce's (2012) contention that it is permissible in these cases for the agent to choose any available act.)

In contrast, it is easy to construct an idealized experimental resolution of *Death in Damascus* that is suitable for studying *decision-theory efficacy*. If Death's reliability is nearly perfect, then nearly all decision-makers end up dead regardless of their decision-making theory. The idealized experimental resolution of *Death in Damascus* suitable for studying decision-theory efficacy, therefore, does not support any theory. However, in the asymmetric variant, all decision-makers end up dead, but some end up dead *and* spend their last day in a less pleasant location. Thus, from the perspective of decision-theory efficacy, we want a theory that recommends the more pleasant location.

Similar points concerning decision-theory efficacy apply to Ahmed's (2014b) variant in which the agent is offered the chance to pay $1 to base their decision on the flip of an indeterministic coin, which Death cannot predict. As Ahmed (2014b: 589) points out, half of those who purchase the coin survive, while nearly all of those who do not die. Therefore, from the perspective of decision-theory efficacy, we want a decision theory that recommends paying.

However, it is less clear that paying is causally *act efficacious*. This is because, at least prima facie, the causal consequences of traveling to Damascus (Aleppo) after viewing the coin are the same as the causal consequences of traveling to Damascus (Aleppo) without viewing the coin—except for the lost $1. Those committed to the idea that paying for the coin is rational may, therefore, need to deny that rationality is determined by causal act efficacy.

A possible response on behalf of CDT would dispute Ahmed's characterization of the alternatives. Ahmed imagines the alternatives to be (i) go to Damascus, (ii) go to Aleppo, or (iii) pay to flip the coin. But according to Joyce (2012), deliberation about whether to choose Damascus or Aleppo results in a "tie," and either is permissible. The agent must, therefore, engage some sort of tiebreaking procedure. Thus, the agent's available alternatives ultimately are (i) engaging a deterministic tiebreaking procedure which outputs Damascus or Aleppo, or (ii) paying $1 to engage an indeterministic tiebreaking procedure which outputs Damascus or Aleppo. CDT recommends (ii) because that has a lower chance of outputting Death's location. The same treatment would then need to be used to amend Lewis's (1981a: 29–30) proposed solution to the Hunter-Richter problem.

to not smoking without cancer, and she prefers smoking with cancer to not smoking with cancer. Should Susan smoke?

It seems clear that Susan should smoke, but *why*? According to nonexperimental approaches to decision theory, the answer is that smoking is more intuitive, or that smoking is supported by a set of axioms that are more intuitive than those that support not smoking. The experimentalist has a better answer.

The description of the case makes clear that Susan takes her current decisions to have no causal effect on the formation of the lesion. Susan also believes that her decision theory has no causal effect (at any point) on the formation of the lesion (otherwise, this would be a classic Newcomb Problem and not a medical one). We incorporate this information into the idealized experimental resolution of the decision problem. In the idealized experimental resolution, the formation of the lesion is determined by an independent process that results in a percentage of the population forming the lesion. (To assume otherwise would be to allow Susan's belief inaccuracy to affect the dependent variable, exactly the sort of effect that we hope to remove in idealized decision theory.) When we analyze the case given this assumption, we infer that adjustments to the decision Susan makes or the decision theory she employs have no causal effect on the development of cancer. Rather, changes to these variables only result in changes to Susan's pleasure in smoking.

Refraining from smoking, and any decision theory which recommends doing so, are thus irrational. Contrary to standard methodology, refraining from smoking is not irrational because it is directly or indirectly unintuitive, but because it is efficacious in preventing smoking pleasure and not cancer. These effects can all be observed, in principle, in the idealized experimental resolution of *The Smoking Lesion*.

2.3.2 Difference between Idealized and Non-idealized Decision Theory

Not all of decision theory employs idealizing conditions. There is research into non-idealized decision theory, typified by the work of Pollock (2006) and Weirich (2004). On the experimental approach, we can neatly distinguish idealized and non-idealized decision theory, and in a way that respects the idea that both types of theories are studying the same concept of rationality.[18]

---

[18] Non-idealized decision theory is often thought to take its inspiration from the "bounded rationality" idea, but this can be misleading. In fact, Pollock and Weirich see themselves as studying decision-making rationality in "realistic" situations, not as studying a different type of rationality. Weirich (2004: 6) explains: "A standard of bounded rationality does not introduce a new

Idealized decision theory attempts to control for the effects of limitations in the agent's epistemic perspective in determining the efficacy of acts or decision theories. Thus, for idealized decision theory, cognitive limitations and inadequate time for calculation represent confounding variables that should be controlled for in an experimental resolution. In contrast, non-idealized decision theory attempts to determine efficacy without controlling for these factors. Thus, for non-idealized decision theory, cognitive limitations and inadequate time for calculation are the conditions under which efficacy is to be determined. In short, on the experimental approach, *both idealized and non-idealized decision theory study efficacy*, but under different experimental resolutions.

For example, when studying the efficacy of decision theories under time limitations, a non-idealized experimental resolution would stipulate a set of decision problems, the cognitive abilities of the agent, and a time limit for calculating decisions. Then, various decision theories would be evaluated as the independent variable. The theories that produce the best weighted average of success – in relation to the decision problems, cognitive ability, and time limit – are the better theories.

When studying the efficacy of acts, rather than decision theories, the distinction between idealized and non-idealized decision theory is less clear. While the weighted average of success produced by a *decision theory* depends on the agent's cognitive abilities and calculation time, that produced by an act does not. An act produces the same results whatever process is used to select it. Therefore, the "best act" for a non-idealized agent with cognitive limitations and time constraints matches the best act for an idealized agent.

The experimental approach to non-idealized decision theory, therefore, focuses on decision theories as the independent variable, not acts. In the next section, we explore reasons to adopt decision theories as the independent variable for idealized decision theory. A positive upshot of doing so would be the creation of an even more unified research program for studying rational decision-making.

3 Decision Theories as the Independent Variable in Idealized Decision Theory

In some contexts, it is best to study the efficacy of acts and in others that of decision theories. In this section, I explain the motivation for thinking of

concept of rationality but rather applies the usual concept in cases where idealizing assumptions are relaxed and agents face obstacles such as a shortage of time for reaching a decision."

decision theories as the appropriate object of study for determining decision-making rationality in idealized decision theory (e.g., in determining whether it is rational to one- or two-box), and I discuss the sorts of theories that follow therefrom.

3.1 Why Study the Efficacy of Decision Theories when Determining Rationality?

Clearly, an agent's physical and cognitive abilities, false beliefs, and bad luck can prevent her from attaining her goals, but it is odd to regard her *rationality* as a further hindrance. Instead, we might think that if we specify the goals and eliminate all differences in abilities, beliefs, and luck, then the decision-makers that tend to do best are those who are most rational. This, anyway, is my construal of the best general motivation behind success-first decision theories, especially those of Gauthier (1986) and McClennen (1990).

Nevertheless, even if it is odd to regard practical rationality as a hindrance, this is exactly what it must be if we identify rational decision-making by determining the efficacy of *acts*. There are many cases, like Newcomb's Problem, where those employing a decision-making theory that recommends the most efficacious acts meet with bad results (in comparison to the results of employing certain other decision-making theories). These cases often involve – as in Newcomb's Problem – predictions of an agent's choices.

One might respond that here the predictor is "rewarding for irrationality." Sometimes the remarks of Gibbard and Harper (1978: 181) and Lewis (1981b: 377) are interpreted in this way. However, I interpret them as claiming merely that the predictor rewards for a predicted one-box decision, which is irrational, rather than claiming that the predictor is "rewarding for" irrationality. The latter claim, after all, is false. The reward condition in Newcomb's Problem is not a prediction of irrationality (a case with this reward condition would be very different from Newcomb's Problem). Rather, the reward condition is a prediction of one-boxing. Since two-boxing is the causally efficacious act, those who think rationality is determined by the causal efficacy of acts should just accept that rational agents end up worse off, without attempting to soften the blow by appealing to the false claim that the predictor's reward condition is irrationality.[19] If, on the contrary, one is

---

[19] Cases involving an infinite number of decisions, such as Arntzenius et al.'s (2004: 262) *Satan's Apple*, provide non-prediction examples in which employing a decision-making theory that recommends always choosing the most causally efficacious acts leads to bad results. Arntzenius

inclined to reject a decision theory because it leads agents to disaster (under idealizing conditions), then the way to do that is to study the efficacy of decision theories.[20] The ultimate justification for this perspective is that decision theories are the proper independent variable.

3.2 Appropriate Experimental Resolutions

As we saw in section 2, a strength of the experimental approach is that it makes clear the value of identifying suitable idealizing conditions when studying idealized decision theory. One obvious idealizing condition is belief accuracy. It would be absurd to demand that idealized decision theories perform well despite belief inaccuracy. This would be like demanding that a flu treatment cure patients when they mistakenly believe that they have the flu. As discussed in section 2, the effects of belief inaccuracy should be controlled for by studying experimental resolutions where agents' credences match objective chances.

The other obvious idealizing condition is unbounded cognitive processing. Idealized decision theory determines what agents should do when they are not facing processing constraints. Accordingly, such constraints should be controlled for in experimental resolutions of decision problems. To include the effects of such constraints in experimental resolutions would be to engage in non-idealized decision theory.

The last restriction on appropriate experimental resolutions that I will discuss is *theory-extensionality*. (I will not refer to this as an "idealizing condition" because it seems to be important for almost all forms of non-idealized decision theory as well.) An experimental resolution is theory-extensional when the conditions under which an agent is rewarded or punished do not essentially reference the decision-making theory that the agent employs. Otherwise, it is theory-intensional. Almost all decision problems

---

et al.'s (2004: 267) suggestion that such agents are being "punished for" their inability to self-bind is just as problematic as the claim that agents in Newcomb's Problem are being punished for their rationality. As Meacham (2010) points out, decision problems are individuated partly by the acts available, and thus cases in which binding acts are available are simply different decision problems. Cases like *Satan's Apple* represent a problem for agents who employ causal act-efficacy decision theories *precisely because* each act on the road to disaster is optimally efficacious. They have this feature *by design*. If, on a metatheoretical level, one cares only about the causal efficacy of acts, then there is no puzzle to be solved: one should grant that such agents are led to disaster.

[20] Presumably, even supporters of focusing on act efficacy like Arntzenius et al. (2004) feel some of this impulse, or they would not attempt to blame the causal decision theorists' disasters on their inability to self-bind.

studied by decision theorists yield theory-extensional experimental resolutions. Even decision problems that seem theory-intensional are actually
theory-extensional. Newcomb's Problem is an example. The reward condition
in the experimental resolution of Newcomb's Problem is a prediction of one-
boxing by a reliable predictor. To say that the predictor is reliable is exactly to
specify a high chance of a one/two-box prediction when one/two-boxing will
in fact occur. What matters for the reward condition is the predicted one-
boxing; the reward condition makes no essential reference to the decision-
making theory that the agent employs.

A theory-intensional resolution would be one where the reward condition
is, for example, *being an evidential decision theorist*. This sort of case is far less
interesting, and is not relevant when determining the efficacy of decision
theories. Theory-intensional resolutions in which agents are punished for
employing certain decision theories are similarly irrelevant when determining
efficacy. For example, the fact that a demon might kill those who employ CDT
is not a mark against the efficacy of the theory. To think otherwise would be
like denying the efficacy of a flu treatment because it is outperformed by a
placebo when all and only the patients receiving the treatment are poisoned.

It remains an open question whether, within idealized decision theory, any
decision theory has optimal efficacy across all appropriate experimental
resolutions. I suspect that there are no such theories, and thus that there is
symmetry between the study of act- and decision-theory efficacy (since it is
also true that no acts are necessarily efficacious). On the experimental
approach, the necessary truths need not concern efficacy but rather *evaluation*. The act experimentalist claims: "necessarily, the rational act is contingently optimally efficacious within the resolution." The decision-theory
experimentalist claims: "necessarily, the rational act is recommended by a
contingently optimally efficacious decision theory within the resolution."

To sum up, there are at least three conditions for appropriate idealized
experimental resolutions of decision problems: belief accuracy, unbounded
cognitive processing, and theory-extensionality. If decision-theory efficacy
determines rationality, then a decision theory is a better guide to rationality
to the extent that it outperforms others relative to the experimental resolutions being studied.

3.3 Success-first Decision Theories

CDT attempts to isolate the causal consequences of acts, while causal success-
first decision theory attempts to isolate the causal consequences of decision

theories. Hypothetical experiments should, in each instance, aim to capture all the effects of the independent variable on the dependent variable, and so the experiment "starts" at a point at which all the effects of the independent variable can be considered.

In the case of act efficacy, this is the point at which an action can be performed, and the potential consequences of available acts are measured from then forward. In imagining the experiment, we hold everything constant that is causally independent of the acts (or evidentially independent, if using the evidential experimental approach discussed in section 2.1) in order to avoid confounding variables. Similarly, for decision-theory efficacy, the experiment starts at the point at which a decision theory can be adopted, and the potential consequences of the available acts are measured from then forward.

According to the causal experimentalist approach, there exist Newcomb-like prediction problems in which it is impossible for the decision theory that an agent employs to causally influence the dependent variable, because, for example, the prediction occurs before the agent was born. When we study experimental resolutions like that, we are not studying decision-theory efficacy, but rather the efficacy of something else. We might, for example, be studying the causal efficacy of an *agent generator*, not of the agent itself. In some contexts, this is a fascinating question, but it is not the study of the resulting agents' decision-making rationality.

In the rest of this section, I evaluate the potential object-level success-first decision theories mentioned in the introduction. Armed with the experimental approach, we will be able to better appreciate the degree to which existing success-first decision theories succeed, and to formulate a plausible way to perfect them.

3.3.1 Gauthier's "Constrained Maximizer"

Gauthier's (1986) "constrained maximizer" is an agent that is disposed to consider not simply which acts maximize expected utility in the present, but also which acts it would have maximized expected utility to plan on choosing in the past. In much of *Morals by Agreement*, Gauthier's discussion exclusively concerns strategic interaction between bargaining partners, but he provides a generalization of his basic idea that covers decision-making more generally. He proposes the following:

> *Gauthier's Proposal*: if at some time t_0 it maximizes expected utility to follow a plan that involves A-ing at some subsequent time t_1, then the agent should A at t_1. (1986, 1989)

I have taken a few liberties with Gauthier's proposal in light of the discussion above. In *Morals by Agreement,* Gauthier's formulation is in terms of an agent's dispositions to act. He claims that if it is rational to adopt a decision-making disposition, then it is rational to subsequently act in accordance with that disposition. Formulating the theory in terms of dispositions raises problems beyond the ones discussed here.[21]

In Newcomb's Problem, there is divergence between the best act in the present and the best plan in the past. At any time before the prediction is made, it would maximize expected utility to follow a one-boxing plan. Therefore, if we set t_0 to a time prior to prediction, Gauthier's proposal would recommend one-boxing. However, two-boxing is recommended if t_0 occurs after the prediction. Thus, if Gauthier's proposal is to give a different recommendation than straightforward maximization in Newcomb's Problem, the point at which plans are evaluated must not be in this post-prediction region. But as it stands, the proposal leaves the point of evaluation for the utility of plans unspecified. We therefore turn to suggestions for specifying this point.

3.3.2 McClennen's "Resolute chooser"

One suggestion is to tie evaluation of the utility of plans to actual commitments or "resolutions" made by the agent. This is McClennen's (1990) proposal. When one of McClennen's 'resolute choosers' determines that plan p is utility-maximizing, she resolves to follow p. When subsequently faced with a decision, the agent ends up "intentionally choosing to act on that resolve" by choosing the act specified by p (McClennen 1990: 15). I suggest that for our purposes it is best to discard references to psychological aspects of "resolve," and to focus on how McClennen's proposal amounts to an alternative to standard decision theory. In this spirit, we can imagine that a "resolution" manifests itself in an agent dropping a temporal anchor when she recognizes a utility-maximizing plan. In the future, when the agent faces a decision, the utility-maximizing status of the action at the time of the anchor dictates what to do. So, in Newcomb's Problem, a resolute chooser who is temporally located before the prediction may see that a one-boxing plan has greater expected utility than a two-boxing plan. She resolves to one-box, thus dropping the

---

[21] For a discussion of some of the problems of the disposition formulation, see Smith 1991. It should be noted that in his later work, Gauthier (1994) retreated from this proposal. His later view is that it is only rational to follow a plan at t_1 if by so acting *one is better off than one would be had one never committed to the plan at all.* This theory retains causal efficacy in Newcomb-like problems that involve "assurances." However, it is causally inefficacious in Newcomb-like problems that involve "threats."

anchor. When the decision arrives, resolute choice recommends one-boxing due to the utility-maximizing status of one-boxing at the time of the anchor.

McClennen's proposal makes clear the past point from which plans are to be evaluated, and it does sometimes create better payouts in Newcomb Problems than those received by causal decision theorists. However, McClennen's resolute choice gives the same recommendations as CDT in situations where agents are not forewarned of predictions. This is the case in the classic Newcomb Problem, where it is commonly assumed that the agent does not know of the prediction until offered the choice. The agent thus has no opportunity to make a resolution, and without a prior resolution, McClennen's theory recommends two-boxing. This is part of a general pattern: decision rules that require an actual pre-commitment will in some cases be less success-conducive than one that allows for merely hypothetical pre-commitments (Andreou 2008). In other words, any theory that requires that an agent *actually* "resolve," "intend," or otherwise "commit" to a plan may fail if the agent is not forewarned of impending predictions.

3.3.3 Meacham's "Cohesive Decision Theorist"

To solve this problem with McClennen's formulation, a theory needs to focus on hypothetical rather than actual commitments. Such a theory is formulated by Meacham (2010: 68–9), in an attempt to advise agents to choose actions that they would have bound themselves to choose. According to Meacham's "cohesive decision theory," agents should make decisions according to the "comprehensive strategy" (a function that maps every decision problem onto one of its available acts) they would choose for themselves from the perspective of their "initial credence function." Meacham imagines that such agents would maximize "cohesive expected utility":

$$\mathrm{CoEU}(CS) = \sum_i \mathrm{ic}(w_i : CS)\mathrm{u}(w_i),$$

where "ic" is the agent's initial credences and "CS" is a comprehensive strategy for selecting acts given decision problems. Meacham uses ":" as a neutral connective between strategies and worlds, leaving open whether this should be understood causally or evidentially. Meacham's formulation of expected utility builds on the standard formulation (i.e., $\mathrm{EU}(A) = \sum_i \mathrm{cr}(w_i:A)\mathrm{u}(w_i)$) by giving control to the agent's initial credences rather than her current credences, and by evaluating plans ("comprehensive strategies") instead of acts. As Meacham (2010: 69 n. 33) notes, we might understand these initial credences as the agent's "ur-priors."

Meacham (2010: 69) explains that a specification of cohesive decision theory with the "standard form" would rely on an agent's current credences about her initial credences:

$$\text{CoEU}(CS) = \sum_i \text{cr}(\text{ic}_i) \sum_j \text{ic}_i\big(w_j : CS\big)\text{u}\big(w_j\big),$$

where "cr(ic)" is the agent's current credences about her initial credences.

Under the experimental approach, we should immediately apply two Newcomb tests when considering the efficacy of cohesive decision theory. The first, as revealed in section 2.1, is that a theory should recommend one-boxing in the classic Newcomb Problem. The second, as revealed in section 2.3.1, is that the theory should recommend the two-boxing equivalent in medical Newcomb Problems. These represent two of the most basic measures when testing a success-first decision theory.

For cohesive decision theory to pass the second test, we must interpret ":" as recommended by CDT. Otherwise, the theory will fail to recommend the two-boxing equivalent in medical Newcomb Problems. Thus, the connection between comprehensive strategies and worlds must be spelled out by the agent's causal dependency hypothesis, and more precisely, the agent's *ur-causal dependency hypothesis* (more precisely still, by the agent's current credences about the agent's ur-causal dependency hypothesis). Doing so will satisfy the medical Newcomb Problem requirement. For example, as we saw in *The Smoking Lesion*, in the idealized experimental resolution, the decision to refrain from smoking is efficacious only in preventing smoking pleasure, not in preventing cancer. The same is true of a nonsmoking plan from the ur-priors. This allows cohesive decision theory to pass the second test, but it prevents it from passing the first.

Appealing to ur-causal dependency hypotheses is a cumbersome way to attempt to produce a one-box recommendation in the classic Newcomb Problem, and it is unclear whether it succeeds. For example, consider a Newcomb's Problem in which the prediction takes place on July 1, 2017 at noon. Now consider which comprehensive strategy would maximize causal expected utility from the ur-priors. If the prediction has already occurred, then a two-boxing strategy is most efficacious; whereas, if the prediction has not occurred, then a one-boxing strategy is most efficacious. Whether it is better to "bind" oneself to a one- or two-boxing strategy thus depends on where the agent is in time – from the ur-priors, the agent asks, "Am I located before or after noon on July 1, 2017?"

Giving control to the ur-priors is like stripping away the agent's evidence, and it is difficult to think about the probabilities an agent would assign to locations in time if she had no evidence whatsoever. However, given that the best strategy for the causalist in Newcomb Problems crucially depends on these probabilities, we must find some way to determine them if cohesive decision theory is to produce a recommendation. These credences seem vague or impossible to specify.[22] This problem might be solved by a new form of success-first decision theory – the "functional decision theory" of Soares and Levinstein (manuscript) and Soares and Yudkowsky (manuscript) – which does not focus on causation and time, but rather on logical counterfactuals regarding the output of the agent's decision function.[23]

To sum up, the experimental approach to studying the efficacy of decision theories provides a backdrop against which to evaluate the success-first decision theories of Gauthier, McClennen, and Meacham. Their analyses go astray, but are moving in the right direction. The justification of a success-first decision theory, I suggested, is not to be located, precisely, in its "cohesiveness," in its being an "asset" rather than a "liability," or related notions. Rather, success-first decision theories succeed exactly when they produce the best average payout in appropriate experimental resolutions of decision problems.

4 Conclusion

This chapter has presented an experimental approach to decision theory, which is compatible with the general aims of both CDT and EDT. When evaluating causal efficacy, we should use decision-theoretic experiments to

---

[22] Even if we grant the possibility of determinate locational probabilities from the ur-priors, the recommendations of cohesive decision theory would still fall far short of an optimally efficacious decision theory. From the ur-priors, presumably, the probability that an agent is located prior to the prediction is the sum of the probabilities of the possible locations prior to noon on July 1, 2017. If the probability of being in this region falls below a certain threshold (e.g., .001 in the classic Newcomb Problem if the predictor has near-perfect reliability), then cohesive decision theory recommends two-boxing. We can lower the probability threshold by imagining that the prediction occurs nearer to the beginning of the total possible days that an agent could be located, or by lessening the difference between the money in the transparent box and the potential reward in the opaque box. These changes affect what cohesive decision theory recommends, even though the causal efficacy of employing a decision theory that recommends one-boxing remains optimal.

[23] The *motivations* for functional decision theory seem to align perfectly with cohesive decision theory. Soares and Yudkowsky (manuscript: 9) write: "Pre-commitment requires foresight and planning, and can require the expenditure of resources—relying on ad-hoc pre-commitments to increase one's expected utility is inelegant, expensive, and impractical...FDT agents simply act as they would have ideally pre-committed to act." Whether functional decision theory makes good on this claim is best judged within the experimentalist framework.

isolate the causal consequences of acts or decision theories. For the causal experimentalist, the disagreement between one- and two-boxing in Newcomb's Problem lies in a disagreement over the appropriate independent variable for decision-theoretic experiments. In line with CDT, two-boxing is supported when acts serve as the independent variable. In line with success-first decision theory, one-boxing is supported when decision theories serve as the independent variable. Several success-first decision theories have been evaluated using the causal experimental approach, and none has been found entirely satisfactory, though the potential for research seems bright. Ultimately, the goal for success-first decision theories is maximization of success in appropriate experimental resolutions of decision problems.[24]

[24] I would like to thank Arif Ahmed, Adam Bales, and Andy Egan for providing detailed comments on earlier drafts of this chapter. I also thank Abram Demski, Matthew Graves, James Joyce, Ben Levinstein, Mary Salvaggio, Nate Soares, Jeremy Strasser, and David Thorstad for helpful comments and discussion.

7 Deliberation and Stability in Newcomb Problems and Pseudo-Newcomb Problems[1]

James M. Joyce

This chapter defends *causal decision theory* (CDT) against some alleged counterexamples that proponents of *Evidential Decision Theory* (EDT) have raised against it. I argue that sophisticated *deliberational* versions of CDT, pioneered by Skyrms (1982, 1990b) and elaborated in Arntzenius (2008) and Joyce (2012), can defuse any of these counterexamples.

The chapter has six sections. Section 1 distinguishes *Newcomb Problems* from *pseudo-Newcomb Problems*. Section 2 addresses predictability and freedom. Section 3 distinguishes CDT and EDT. Section 4 defends CDT's handling of Newcomb Problems. Section 5 introduces the notion of deliberational equilibrium, and distinguishes picking from choosing. Section 6 considers "unstable" decisions and defuses counterexamples from Spencer and Wells (2017) and Ahmed (2014b).

1 Newcomb Problems

In *Newcomb Problems*, choices correlate with features of the world that choosers cannot causally influence. As a result, acts that *cause* desirable/undesirable future results can also *indicate* undesirable/desirable past events, leaving agents to wonder whether to causally promote desirable outcomes or to produce news of desirable outcomes they do not control.

All Newcomb Problems can be subsumed under a common rubric. Imagine an idealized agent, let's say *you*, who is now (time t_1) facing a decision, and a predictor, *Omega*, who at *past* time t_0 made a guess about how you would act on the basis of an examination of your t_0 brain state (a common cause of both his guess and your choice). Things are arranged so that the outcome of your

[1] This chapter benefited greatly from discussions with Arif Ahmed, Brad Armendt, Kevin Blackwell, Simon Huttegger, Sarah Moss, Huw Price and Brian Skyrms. An earlier draft of this chapter, with additional relevant material, may be found at https://deepblue.lib.umich.edu/documents. All citations should be to the published version.

choice depends on Omega's guess, but your only evidence about this comes via your knowledge of your own current beliefs and desires, wich are clues to your t_0 brain state.

For our purposes, any (ideal) Newcomb Problem satisfies the following:

NP_1 For each possible act A there is an associated "type-A" brain state, and you are subjectively certain that you will do A at t_1 iff you occupied the type-A state at t_0, hereafter A^{T}.[2] You typically will *not* know your type until you know what you (irrevocably) choose.

NP_2 Omega tried to discern your type at t_0, and guessed that you would choose A, hereafter A^{π}, iff he identified you as type-A. Omega may or may not be a perfect identifier of types. But, for all acts A and B, you know Omega's chances of misclassifying you as type-B when you are really type-A.

NP_3 The past is *fixed*. You cannot now change your t_0 state, Omega's prediction, or any other past fact.

NP_4 Omega is *reliable*, i.e., better than chance at predicting your act. Moreover, absent further evidence, your confidence in Omega's reliability is constant throughout the decision-making process.

NP_5 You are *free*. No obstacles prevent you from choosing whichever act you ultimately judge best. Crucially, whatever Omega predicted and however reliable he is, you have the power to falsify his prediction. You might not want to exercise this power and might be sure that you will not, but you can.

I take NP_1-NP_5, as elaborated above, as *definitive* of Newcomb Problems. People frequently mistake decisions satisfying only some of these conditions for genuine Newcomb Problems, and wrongly portray solutions to such *pseudo*-Newcomb Problems as answers to the real thing.

To explain NP_1-NP_5 it helps to have a formal model of you as a decision maker.[3] You face a free choice among *acts* $\{A_1, A_2, \ldots, A_M\}$ whose *outcomes* depend on which member of a partition $\{S_1, S_2, \ldots, S_N\}$ of *states* obtains. Each act/state pair fixes an outcome $O_{m,n}$ that encompasses all relevant consequences of A_m when S_n obtains. States describe features of the world that you cannot influence, but which may affect the outcome of your act. In light of NP_3,

---

[2] We could have let you be merely highly confident about type-act correlations without altering the discussion.

[3] This is admittedly an idealization, but in all the cases we consider real agents approximate the behavior of their ideal counterparts.

each S_n will specify both your type and Omega's guess, i.e., each is a conjunction $A_j{}^\top$ & $A_i{}^\pi$ & . . ., the ellipsis capturing any further facts that affect outcomes.

At each time t you are endowed with a subjective utility function $util_t$ that measures the desirabilities of outcomes, and a *credence* function $prob_t$ that encodes your degrees of belief in events that might influence the outcome of your choice. Since we will not consider changes in desire, $util_t$ is assumed constant. There is some dispute over the proper domain of $prob_t$, but all agree that probabilities of states *conditional on acts* are well-defined. So, you always have a definite estimate, $prob_t(S_n/A_m)$, of the probability that S_n will obtain if you do A_m.[4]

To illustrate, consider first the *Flagship* Newcomb Problem: you choose between act One of taking an opaque box that contains \$1,000,000 iff Omega predicted that you would take *only* that box, and act Two of taking the opaque box plus a transparent box containing \$1000. We represent your decision thus (\$1000 = 1 utile):

Table 1

Flagship	One$^\top$ & One$^\pi$	One$^\top$ & Two$^\pi$	Two$^\top$ & One$^\pi$	Two$^\top$ & Two$^\pi$
One	1000, p	0, $1 - p$	1000, 0	0, 0
Two	1001, 0	1, 0	1001, $1 - q$	1, q

Acts are at the left, states across the top. The first entry in each cell is the utility of the outcome received there. The second is your estimate of the probability of the cell's associated state *conditional on its act*. Your estimates of the probabilities that Omega correctly guessed your choices of One or Two are $p = prob(\text{One}^\pi/\text{One})$ and $q = prob(\text{Two}^\pi/\text{Two})$, which both exceed 0.5 (**NP₄**). Both are one when Omega is a *perfect predictor*.

Gibbard and Harper's (1978) *Death in Damascus* provides another example. You are on the highway between Damascus and Aleppo, and will suffer a fate worse than death if you fail to arrive in one of the cities by nightfall. Alas, your prospects are little better if you do arrive. Yesterday Omega (irrevocably) predicted your destination, and sent assassins to the

---

[4] **NP₁** requires $prob_t(S_n/A_m) = 0$ whenever S_n entails $A_k{}^\top$ for $k \neq m$. **NP₄** requires that the sum of the $prob_t(S_n/A_m)$ for which S_n entails $A_m{}^\pi$ (i.e., the probability that Omega correctly guessed A_m) exceeds ½.

predicted city. To go there is to die; to go the other way is to live. With $1 = $ life and $0 = $ death, your decision is:

Table 2

DD	Alep^τ & Alep^π	Alep^τ & Dam^π	Dam^τ & Alep^π	Dam^τ & Dam^π
Alep	0, p	1, 1 − p	0, 0	1, 0
Dam	1, 0	0, 0	1, 1 − q	0, q

Here you have a mortality rate of $p = prob(\mathrm{Alep}^\pi/\mathrm{Alep})$ or $q = prob(\mathrm{Dam}^\pi/\mathrm{Dam})$ depending on whether you choose Aleppo or Damascus. Again, we can imagine Omega as perfect, in which case your mortality rate is 1 wherever you go.

We discuss these examples in detail below, but first let's better understand NP_1-NP_5.

2 Reconciling NP_1-NP_5: How You Can Be Free while Omega Is Reliable?

Newcomb Problems asks us to square three seemingly irreconcilable beliefs. You, the agent, must be convinced that (a) you cannot affect Omega's prediction, (b) Omega *reliably* predicts your choices, but (c) you are free to falsify his prediction. But, how can you be sure that Omega's prediction and your choice will coincide unless one affects the other?

Let's start by asking how you can know you will choose A at t_1 iff you occupied the type-A brain state at t_0, as NP_1 requires. The answer is that, you *count* as type-A at t_0 when, given your *actual* circumstances at t_0 and after, you end up disposed to rank A among your best options at t_1 and to pick it over similarly ranked options. Dependence on *actual* circumstances is crucial. As a type-A you will choose A in the actual world, but need *not* do so in other possible worlds (though "type-A" denotes a different brain state there). A two-box type in Flagship will actually choose Two (probably because it dominates One). But, if something had disturbed its reasoning between t_0 and t_1, it might have chosen One. Thus, there is only a *contingent* connection between being in the type-A state at t_0 and choosing A at t_1. Differently put, *intrinsic* features of the type-A state (which "being type-A" is *not*) could cause different choices at t_1 depending on what transpires between t_0 and t_1.

The upshot is that your act is predictable from intrinsic features of your t_0 brain state only insofar as your actual situation is known. This helps us understand NP_2. Omega may know enough about your situation to deduce a definite t_1 choice from each intrinsically described t_0 brain state. Unless he is perfect, however, he will not know which intrinsic t_0 state you occupy, and so might misidentify your type. You, in contrast, might know more than Omega does about your t_0 brain state, but you might be unable to identify your type before t_1 because you lack information about the situation between t_0 and t_1. So, you can be certain you will choose A at t_1 iff you are type-A at t_0, and yet not know your type until you (irrevocably) decide what to do.

NP_3 requires you to regard the past as *causally* independent of your act. You must be confident that, whichever past is actual and however you actually choose, that same past would have obtained had you chosen otherwise. Following Stalnaker (1972), Gibbard and Harper (1978) we characterize this requirement of *causal act/state independence* using subjunctive conditionals: you regard S as causally independent of A just when your credences for $A \,\square\!\!\rightarrow S$ and S are equal.[5] This gives us:

> NP_3 (*Causal Independence*) For each state S and acts A and B, you are certain that if $A \,\&\, S$ actually holds, then S would still have held had you chosen B, so that $prob_t(B \,\square\!\!\rightarrow S \,/\, A \,\&\, S) = 1$ at every t.

This makes states counterfactually independent of acts: $prob_t(A \,\square\!\!\rightarrow S) = prob_t(S)$.

NP_3 requires you to regard the past as beyond your present influence. You must believe that acting differently would not alter your type or Omega's prediction, so that, for any A, B, C, and D, $prob_t(B \,\square\!\!\rightarrow S \,/\, C^\tau \,\&\, D^\pi) = prob_t(B \,\square\!\!\rightarrow S \,/\, A \,\&\, C^\tau \,\&\, D^\pi)$. In Flagship, you must be convinced that if you take only the opaque box and get the million, then you would still have gotten the million had you taken both. In DD, you must believe that if you are in fact going to die in Aleppo, then you would have lived had you fled to Damascus. *You are not in a Newcomb Problem if you lack these sorts of beliefs!*

Turning to NP_4, we need to know what it means to regard Omega as *reliable*. Does it mean that he is likely to be correct *given what he guessed*, or *given what you choose*? It is the second notion that matters. Seeing yourself as free is consistent with confidence (even certainty) that Omega has correctly

---

[5] This is only the simplest of many ways of characterizing causal act/state independence. For more general approaches, see Skyrms (1980); Lewis (1981a); Meek and Glymour (1994); Pearl (2000); Joyce (2010a).

predicted *the act you will actually do*, but not with confidence that you will do *whatever he predicted*. In Flagship, you assign high credence to both these "prediction-if-act" conditionals:

> If I do choose One/Two, then Omega is likely to have predicted One/Two.
> But, you assign low credence to at least one of these "act-given-prediction" conditionals:
> If Omega predicted One/Two, then I am likely to choose One/Two.

To the extent that you see yourself as free, your credences for these latter conditionals coincide with your credences for their *consequents*, which rules out being confident in both. Suppose you come to see One as your best option. As a free agent you will *not* reason like this: "One is my best option, but I won't choose it if Omega predicted Two, because he is so reliable." Rather, you reason: "I'll choose my best option whatever Omega predicted. So, I'll choose One even if he predicted Two. But, since One is my best option, Omega probably predicted One." As a free agent you are confident that you will do what you most prefer, whatever Omega predicted, and however reliable you take him to be. But, you are also confident that Omega predicted the act you will most prefer.

This brings us to NP_5, and what it means to be free. The relevant notion is familiar from compatibilist theories of free will. You see yourself as free when you are confident that (i) no constraints prevent you from choosing an act that you judge to be among your best options, and (ii) your choice is the immediate effect of your ranking the chosen act among your best options. In Newcomb Problems, (i) is the idea that your choice is not constrained by your past state or by Omega's prediction. Even conditional on A^π & B^τ, the *only* thing that prevents you from choosing any act on the menu is that you do not rank it among your best options. In Flagship, nothing prevents one-box types from choosing Two, or two-box types from choosing One. Agents eschew "contrary-to-type" choices not because they *cannot* make them, but because they do not see making them as being in their interests.

Clause (ii) requires your credences for actions to respond to evidence in a distinctive way. You should be confident of performing A only if you rank it among your best options. To a first approximation, other considerations should affect your estimate of A's probability only insofar as they convey information about A's merits relative to other acts. I will say more later, but the salient point now is that, when freely choosing, your beliefs about acts are driven *exclusively* by your judgments about which of them best satisfy your desires, and your belief that you will choose an option that you ultimately regard as best.

3 Causalism and Evidentialism

In Newcomb Problems, acts influence the future and indicate the past. By performing A you bring about outcomes that A causes and create evidence for thinking that you occupied the "type-A" state at t_0. Choosing One in Flagship secures you whatever is in the opaque box, and creates evidence that it is not empty. Choosing Alep in DD causes you to be in Aleppo, and creates evidence that Omega's assassins are already there. Here is the question that divides CDT and EDT: Should you choose acts exclusively on the basis of what they cause or also on the basis of what they non-causally indicate? CDT says that only causal consequences matter; EDT considers purely evidential implications as well.

This disparity emerges in calculations of expected utility. The theories agree that an act's value is a probability-weighted average of the utilities of its potential outcomes, but CDT weights each outcome by the *unconditional* probability of the state that brings it about, while EDT weights it by the probability of that state *conditional on the act. Under appropriate conditions* (see below), A's choiceworthiness in CDT is given by its *efficacy* value $U(A_m) = \Sigma_n prob(S_n) \cdot util(O_{m,n})$, while A's choiceworthiness in EDT is its *news value* $V(A_m) = \Sigma_n prob(S_n/A_m) \cdot util(O_{m,n})$. Differences in U-values reflect expected disparities in desirabilities of future outcomes that acts *cause*, while differences in V-values reflect expected disparities in evidence that acts provide about future *or past* facts.

Causation and indication, which usually go together, diverge in Newcomb Problems. Causalists look at Flagship like this, where x is your credence for being a one-box type, $p = prob(\text{One}^\pi|\text{One})$ and $q = prob(\text{Two}^\pi|\text{Two})$:

Table 3

Flagship	One$^\top$ & One$^\pi$	One$^\top$ & Two$^\pi$	Two$^\top$ & One$^\pi$	Two$^\top$ & Two$^\pi$
One	1000, px	0, $(1 - p)x$	1000, 0	0, 0
Two	1001, 0	1, 0	1001, $(1 - q)(1 - x)$	1, $q(1 - x)$

The sum of the second entries in each column is the unconditional probability of that column, and the expected utilities are:

$$U(\text{One}) = 1000 \cdot px + 0 \cdot (1 - p)x + 1000 \cdot (1 - q)(1 - x)$$
$$+ 0 \cdot q(1 - x)$$
$$= 1000 \cdot (px + (1 - q)(1 - x)),$$

$$U(\text{Two}) = 1001{\cdot}px + 1{\cdot}1 - p)(x + 1001{\cdot}(1 - q)(1 - x)$$
$$+1{\cdot}q(1 - x)$$
$$= U(\text{One}) + 1.$$

Thus, CDT favors Two over One. EDT computes expected utilities like this:

$$V(\text{One}) = 1000{\cdot}p + 0{\cdot}(1 - p) + 1000{\cdot}0 + 0{\cdot}0 = 1000{\cdot}p,$$

$$V(\text{Two}) = 1001{\cdot}0 + 1{\cdot}0 + 1001{\cdot}(1 - q) + 1{\cdot}q = 1001 - 1000{\cdot}q.$$

This favors One over Two when Omega is sufficiently reliable ($p + q > {}^{1001}/_{1000}$).

Death in Damascus is more complicated. Here is CDT's picture, x being your credence in being an Aleppo type:

Table 4

DD	Alep^{T} & Alep^{π}	Alep^{T} & Dam^{π}	Dam^{T} & Alep^{π}	Dam^{T} & Dam^{π}
Alep	0, px	1, $(1 - p)x$	0, 0	1, 0
Dam	1, 0	0, 0	1, $(1 - q)(1 - x)$	0, $q(1 - x)$

Then, $U(\text{Alep}) = (1 - p)x + q(1 - x) = 1 - U(\text{Dam})$, and $U(\text{Alep}) \geq U(\text{Dam})$ iff $x \leq (q - \frac{1}{2})/[(p - \frac{1}{2}) + (q - \frac{1}{2})]$. The evidential utilities are $V(\text{Alep}) = 1 - p$ and $V(\text{Dam}) = 1 - q$, and $V(\text{Alep}) \geq V(\text{Dam})$ iff $q \geq p$. So, EDT recommends whichever city is associated with the type that Omega has the hardest time predicting *irrespective of your beliefs about your type*. CDT's recommendation, in contrast, depends on how confident you are about your type. Notice that, in addition to ranking acts using different criteria, U's values are higher than V's. This is because U, with its focus on the future, factors out the bad news of having to face DD in the first place, while V reflects this.

CDT and EDT also differ about the values of decisions as *wholes*. In any decision your time-t credences are a snapshot of your views about what you are likely to do and what then occurs. You can use these credences to attach a utility to your overall predicament, the *status quo* as Skyrms (1990b) calls it, by averaging expected utilities of acts, weighting each by its probability. In CDT, $U(SQ) = \Sigma_m \, prob(A_m){\cdot}U(A_m) = \Sigma_m\Sigma_n \, prob(A_m){\cdot}prob(S_n){\cdot}U(O_{m,n})$ is your best estimate of the improvement/decline in your fortunes that will occur as an effect of making the decision. In EDT, $V(SQ) = \Sigma_m \, prob(A_m){\cdot}V(A_m) = \Sigma_m\Sigma_n \, prob(S_n \, \& \, A_m){\cdot}U(O_{m,n})$ is the news value of making the decision. In DD, $U(SQ) = 1 - x + (2x - 1){\cdot}U(\text{Alep})$, while $V(SQ) = 1 - (xp + (1 - x)q)$. At $x = {}^{1}/_{3}$,

$p = 0.9$ and $q = 0.7$ the numbers work out so that $U(SQ) = 0.5$ and $V(SQ) = 0.233$. CDT is thus indifferent between facing DD and playing Russian roulette with three bullets in a six-shooter, while EDT ranks DD below playing with four bullets. This difference is to be expected since CDT ignores the "bad news" of having to play Russian roulette in the first place.

Notice too that CDT assesses decisions differently when considering them *in prospect* than when they are actively being made. You consider a choice among acts $A_1, \ldots, A_M$ in prospect when you will choose among them at some *future* time, but cannot choose *now*.[6] Viewed in prospect, acts in future decisions are treated not as current *options*, but as potential *outcomes* lying causally downstream of your current choice. And, as with anything not under your *current* control, CDT assesses future acts using their current *news values*. For example, if you can decide now whether to face DD *tomorrow*, then Alep is not a current option; it is a potential consequence whose current value is $V(\text{Alep})$. This value is very low: learning that you will choose Alep is bad news, now and tomorrow. Of course, CDT says that tomorrow you should not worry about bad news that you *cannot* then control (e.g., being caught in DD), but only about things that you can control. But, it also says that you should consider Alep's news value when deciding whether to expose yourself to a future decision in which it might be chosen. For example, when deciding whether to face DD, the fact that facing it is a strong indicator of your death is highly relevant. Thus, although CDT values a decision you can now make at $U(SQ)$, it values the same decision in prospect at $V(SQ)$. As a result, CDT and EDT often agree about which future decisions to make even if not about how to make them.

4 Pseudo-flagship Fallacies

Opponents of CDT press two sorts of objections. Some, like Levi (2000), claim that the theory is incoherent, e.g., because it lets agents have credences for acts. Others, like Horgan (1981), Egan (2007), and Ahmed (2014a, 2014b), claim that CDT gives bad advice. I have discussed the first worry elsewhere;[7] here I focus on the second. First, I consider arguments against two-boxing in Flagship. Then, after elaborating CDT in section 5, I address more complicated

---

[6] You cannot "tie yourself to the mast" and irrevocably commit to performing these future acts. That would amount to making the future decision now.

[7] See Joyce 2002, 2007; Rabinowicz 2002.

cases in section 6. A unifying theme is that many objections to CDT involve a kind of bait-and-switch in which answers to a *pseudo*-Newcomb are presented as answers to true Newcomb Problems.

For instance, suppose you are going to face Flagship at time t_1 with $p = 0.8$ and $q = 0.9$. But, at $t_{-1} < t_0$ (*before* Omega guesses) you can choose your *type*, and Omega will base his guess on what you choose. There is no reneging: if you choose to be a one/two-box type at t_{-1}, you will freely choose One/Two at t_1 (because you will want to). Clearly, you should choose One$^\mathrm{T}$. This will cost a utile at t_1, but that's cheap for an 80% chance at 1000 utiles. But, if you choose to be a two-box type at t_{-1}, you get the extra 1 utile, but only a 10% chance of an added 1000. Here everyone recommends choosing to be a one-box type at t_{-1}: it maximizes *both U* and *V*.

Some think it incoherent for CDT to recommend One$^\mathrm{T}$ over Two$^\mathrm{T}$ at t_{-1}, but Two over One at t_1. This is sometimes portrayed as an unwillingness of CDT to stand behind its advice. If Two is right at t_1, shouldn't you strive to be the type who chooses Two then? Others see temporal inconsistency. By recommending One$^\mathrm{T}$ at t_{-1} doesn't CDT implicitly endorse One at t_1, and then reverse itself and endorse Two at t_1? Still others (e.g., Yudkowsky 2010) see "reflective incoherence": agents who maximize *U* at all times will wish at t_1 that they had used a decision rule that chose One at t_{-1}.

These worries presuppose that by choosing One$^\mathrm{T}$ in One$^\mathrm{T}$-vs-Two$^\mathrm{T}$ you somehow sanction One in One-vs-Two. This is mistaken. You make the first choice *before* Omega's guess, when you can still influence it. You make the second choice when Omega's guess is part of the inalterable past. But, CDT's advice is consistent. It *always* says that, at any time t, choose *from among the options available to you at t* an act that you expect to be most efficacious at causing desirable results. CDT only seems inconsistent when we forget that One$^\mathrm{T}$-vs-Two$^\mathrm{T}$ and One-vs-Two involve different options with different causal properties exercised at different times. In One$^\mathrm{T}$-vs-Two$^\mathrm{T}$ you can influence whether you get the million, and CDT recommends the action, from among those available at t_{-1}, that is most likely to bring this about. That's One$^\mathrm{T}$. But, in One-vs-Two your type is fixed, and CDT tells you to cause the best results given your t_1 options. That's Two.

Even though Two is what you *should* do at t_1, it is not what you *will* do if you chose *rationally* at t_{-1}. If you chose One$^\mathrm{T}$ at t_{-1} then, while you are free to choose Two at t_1, you will not, because, as a one-box type, you mistakenly favor One. You might favor One because your t_{-1} choice made you an EDTist, or misled you into thinking that you can alter the past, or maybe clouded your cerebellum with vapors of black bile. Who cares? The point is that sanctioning

One^T over Two^T is no endorsement of One over Two. It is no part of your goal at t_{-1} to choose rationally at t_1. Your only goal at t_{-1} is to choose *from among the options available to you then* the act that you expect to best promote desirable outcomes. From CDT's perspective, your t_{-1} options are "One^T-and-irrationally-choose-One" and "Two^T-and-rationally-choose Two." Since the former causes the best outcome, CDT recommends it even at the cost of later irrationality. The moral is that *at all times* CDT endorses the same choice in each decision: One^T over Two^T at t_{-1}, Two over One at t_1. Even though the first choice makes you botch the second, the 1 utile penalty for irrationality at t_1 is more than offset by the 70% increase in your chance at 1000 utiles you get by acting rationally at t_{-1}. Moreover, you will never wish that you had used a decision rule other than CDT. You might wish that you had different options (e.g., One^T-at-t_{-1}-and-Two-at-t_1), but that involves no reflective inconsistency.

5 Decision Instability

Despite its centrality in the literature, Flagship is not a typical Newcomb Problem. Since two-boxing *dominates* one-boxing, CDT's recommendation does not depend on your credences. In problems like DD, credences matter. In particular, information about how likely you are to perform various acts can be evidence about both what future outcomes those acts might cause and what past facts they non-causally indicate. While CDT regards the latter as irrelevant, it *requires* you to consider the first sort of data. This is why I said CDT requires U-maximization "under appropriate conditions." Properly understood, it has you maximize U *only after taking into account all readily available evidence about what your acts may cause.*

While I will not reargue it here, Joyce (2012) contends that you have not processed all your evidence about what your acts might cause until your credences and expected utilities reach a *deliberational equilibrium* ($prob^*$, U^*) in which every act of positive probability has the utility of the status quo. Here I follow a trail blazed by Skyrms (1982) and travelled by Arntzenius (2008) by modeling deliberation as an information-gathering process in which you learn the best ways to pursue desirable outcomes by comparing the efficacies of acts to that of the status quo. Acts with U_t-values higher/lower than $U_t(SQ)$ are seen as better/worse than average at promoting your aims. In general, you want acts with U_t-values exceeding $U_t(SQ)$ to have their credences increased at the expense of acts with U_t-values below $U_t(SQ)$. Yet, you cannot alter credences *ad lib*. Like any beliefs, beliefs about your acts should only change in response to evidence. But, not just any evidence. Since you see yourself as free to choose whichever act you deem best, in deliberation your credences for acts should respond only to evidence about their choiceworthiness. Other factors may affect act probabilities *only* by affecting your views about choiceworthiness. The evidential relations work this way during deliberation because (a) you see yourself as a free agent who will do what you ultimately prefer, and (b) you treat the fact that $U_t(A)$ exceeds $U_t(SQ)$ as (inconclusive) evidence that A will rank among your most preferred options when all the evidence is in.

The details of the deliberative process are not critical, but the idea is that, at each time t, you acquire information about the efficacies of options by learning a conjunction $[U_t(SQ) = u \ \&_m \ U_t(A_m) = u_m]$. But, you should not *choose* on the basis of these U-values until you know they incorporate all readily available information about what your acts might cause.[8] This happens when $U_t(A_m) = U_t(SQ)$ for all A_m with $prob_t(A_m) > 0$. If $prob_t$ and U_t pass this test, then you have achieved an equilibrium and all relevant evidence has been processed. If not, you must update using a belief revision rule that *seeks the good* (Skyrms) by mapping your time-t credences and utilities onto time-$t+1$ credences and utilities in such a way that $prob_{t+1}(A) \geq prob_t(A)$ iff $U_t(A) \geq U_t(SQ)$. For this purpose, I like *Bayesian dynamics*, which sets $prob_{t+1}(A) = prob_t(A) \cdot [U_t(A)/U_t(SQ)]$.[9] Once act probabilities are updated, all other credences are revised via a *Jeffrey shift*: $prob_{t+1}(\bullet) = \Sigma_m \ prob_{t+1}(A_m) \cdot prob_t(\bullet/A_m)$. Since this shift satisfies $prob_{t+1}(S/A) = prob_t(S/A)$, it disturbs neither your confidence in Omega's reliability nor your news values. Note also that this process adjusts act credences *only* in response to evidence about U_t-values. This is critical: in the midst of deliberation, a rational agent's beliefs about acts change only in response to evidence about the merits of those acts.

In all cases that we consider a *deliberational equilibrium (prob*, U*)* will eventually be reached, and updating on U^*-values has no further effect on your credences. You are left with a set of *live acts* $B = \{B_1, \ldots, B_K\}$ such that $prob^*(B_k) > 0$, $\Sigma_k \ prob^*(B_k) = 1$, and $U^*(B_k) = U^*(SQ) \geq U^*(A_m)$ for

---

[8] Even if there are modest costs for acquiring evidence, my conclusions still hold. Also, if you *must* choose before you have time to process all relevant causal information, then CDT tells you to maximize efficacy value relative to your current, imperfect beliefs. CDT can say this while still insisting that you would have made a better decision if you had more time to gather information. Thanks to Brad Armendt for pressing me on this.

[9] Utilities are measured on a positive scale. Utilities can always be scaled this way for decisions with only finitely many acts, or where utilities are bounded.

any A_m. Any act not among the B_k is moot since you are sure you will not choose it.

When B contains only one act, this is what CDT mandates. Any version of Flagship has a unique equilibrium in which you are certain you will take two boxes, confident (or certain) that Omega guessed this, and expecting to get far less than 1000 *whatever you do*.

When multiple acts survive in equilibrium, CDT is indifferent among them. This happens in DD. If p and q both exceed one-half, DD has a unique equilibrium with $0 < prob^*(\text{Alep}) = (q - \frac{1}{2})/[(p - \frac{1}{2}) + (q - \frac{1}{2})] < 1$ and $U^*(\text{Alep}) = U^*(\text{Dam})$. In such cases, you must *pick*. "Pick" is a term of art for a choice process which selects one from a set of equally good acts in a way that is *not* sensitive to differences in utility. (Think Buridan's ass!) Picking is inherently arational. Picking A over B does *not* imply that you have more *reason* to choose A than B.

Your picking method is a fact about your "type" that goes beyond those features of your t_0 mental state that affect which acts you deem choiceworthy at t_1. For simplicity, I assume your type *determines* your pick.[10] This means that an Aleppo/Damascus-type is someone who picks Alep|Dam in the DD equilibrium. I want to emphasize that you will not *care* how you pick. By arriving at equilibrium, you ensure that you see every live act as maximally efficacious in light of all available evidence about what your acts might cause. As far as *desires* are concerned, a pick is an irrelevant detail.

Even so, you will have *beliefs* about your picking tendencies, and these explain why Omega is better than you at predicting your choices: he has better information about how you'll pick! At t_0 he learns two sorts of facts about type: facts that help him deduce the equilibrium that you will settle at, and facts about how you will pick once there. If his evidence about these things is better than yours, he will be better at predicting your behavior. And, his evidence is better. In equilibrium, your confidence that you will pick A is $prob^*(A)$. As stressed above, this is sensitive only to evidence about U-values, and in equilibrium you have taken all such data into account. Thus, $prob^*(A)$ is your fully informed estimate of the probability that you are the "pick-A" type. If you think Omega has better information about your type than you do, then you will expect him to better predict your picks.

---

[10] In a fuller treatment, your type would determine your *chances* of picking various ways. Your credence for A is then your expectation of that chance, so $prob^*(A) = \int_M prob^*(M) M(A) dM$ where M ranges over picking mechanisms, and $M(A)$ is your chance of picking A when mechanism M is in effect.

6 Death and Damascus, and Some Variants

Many of CDT's detractors allege that it mishandles decisions in which multiple acts survive into equilibrium. These are mostly bait-and-switch arguments, albeit of a subtler variety than those already encountered.

Consider first a version of DD where $p = prob(Alep^\pi/Alep)$ exceeds $q = prob(Dam^\pi/Dam)$, say $p = 0.9$ and $q = 0.7$. Here the unique equilibrium is $prob(Alep) = 1/3$. So, if CDT is correct, not only is it acceptable to pick Aleppo, you should have a one-in-three probability that you will. Doesn't that seem wrong, given that 10% of those who go to Aleppo survive, while 30% of those who go to Damascus survive? Shouldn't you choose a higher survival rate (as EDT says), and be certain you will?

Definitely! But, that's *not* your choice. This is the "choose your type" fallacy in new garb. Your survival rate in DD causally depends on your choice *and your type*. If you are an Aleppo-type, you choose between *Aleppo-and-dying-with-probability-0.9* and *Damascus-and-dying-with-probability-0.1* (*not 0.7*). If you are a Damascus-type, you choose between *Damascus-and-dying-with-probability-0.7* and *Aleppo-and-dying-with-probability-0.3* (*not 0.9*). Unfortunately, being unsure of your type, you cannot know which decision you face. Having arrived at the $prob^*(Alep) = 1/3$ equilibrium, you have acquired as much data as you can, but the decision's diabolical structure – with one city offering better survival rates as the other grows more likely – makes it impossible to know which decision you face *until after you pick*. Before you pick, your credence is 1/3 that you face {Aleppo & 0.9 death, Damascus & 0.1 death} and 2/3 that you face {Aleppo & 0.3 death, Damascus & 0.7 death}. Picking resolves the uncertainty. To their dismay, Aleppo-types learn that they have chosen a 90% chance of death in Aleppo, when they could have had a 90% chance of life in Damascus. Damascus-types are slightly less distressed to learn that they opted for a 70% chance of death in Damascus over a 70% chance of life in Aleppo. But nobody chooses from {Aleppo & 90% death, Damascus & 70% death}. To do that they would have to choose their type, which NP_3 prohibits.

Going to Damascus is optimal if you can choose your type *before Omega guesses*. You will then be choosing between these options:

- $Alep^\top$ now and choose from {Aleppo & 90% death, Damascus & 10% death} later on when you see Aleppo as the better option.
- $Dam^\top$ now and choose from {Aleppo & 30% death, Damascus & 70% death} later on when you see Damascus as the better option.

The second option is best. While choosing it will cause your future self to choose the worse future option, it secures you an extra 20% survival probability. This is an exact analog of the "choosing to be a one-box type" decision. As before, it does nothing to undermine CDT.

The idea that you choose your survival rate in DD is hard to shake, as a recent paper by Jack Spencer and Ian Wells (2017) illustrates. Spencer and Wells offer a counterexample, *The Semi-Frustrater*, which allegedly undermines CDT's dominance principle. Retelling their story as a version of DD, suppose you (irrevocably) choose Aleppo or Damascus by pointing toward the chosen city with your right or left hand. The twist is that Omega is better at predicting righties than lefties, but righties get cake (0.05 utiles)! Your situation looks like this, where w, x, y, and $z = 1 - (w + x + y)$ is your credence for the act in its associated row, and $p_R > p_L$ and $q_R > q_L$:

Table 5

SF	Alep$^\text{T}$ & Alep$^\pi$	Alep$^\text{T}$ & Dam$^\pi$	Dam$^\text{T}$ & Alep$^\pi$	Dam$^\text{T}$ & Dam$^\pi$
Alep$_R$	0.05, $w \cdot p_R$	1.05, $w \cdot (1 - p_R)$	0.05, 0	1.05, 0
Alep$_L$	0, $x \cdot p_L$	1, $x \cdot (1 - p_L)$	0, 0	1, 0
Dam$_R$	1.05, 0	0.05, 0	1.05, $y \cdot (1 - q_R)$	0.05, $y \cdot q_R$
Dam$_L$	1, 0	0, 0	1, $z \cdot (1 - q_L)$	0, $z \cdot q_L$

For any initial beliefs that give both righty acts positive credence, SF has the same equilibrium as DD, so that $prob^*(\text{Alep}_R) = (\frac{1}{2} - q_R)/[(\frac{1}{2} - p_R) + (\frac{1}{2} - q_R)] = 1 - prob^*(\text{Dam}_R)$.

Lefty acts are thus inconsequential. Their initial probabilities are moot, and it does not matter how reliably Omega predicts them. Under any conditions, their equilibrium probabilities vanish because some "cake" option always has higher equilibrium expected utility.

Spencer and Wells see this as wrong. They claim that rationality requires you to use your left hand, and permits choosing either city that way.[11] "Consistent right-handers," they write, "end up poorer than do consistent left-handers because they choose irrationally" (p. 11). This remark occurs as part of a discussion of the *Why Ain'cha Rich* (WAR) argument in which Spencer and Wells argue that, while WAR fails to justify one-boxing in Flagship, it does justify left-handing in SF. Echoing a well-known line, they

---

[11] Here Spencer and Wells assume $prob(\text{ALEP}^\pi/\text{ALEP}_L) = prob(\text{DAM}^\pi/\text{DAM}_L)$.

correctly argue that the much ballyhooed fact that one-boxers end up richer than two-boxers cuts no ice against CDT because, through no merit of their own, one-box types *start out* with better options. They cannot help being rich no matter how poorly they choose, while two-box types cannot help being poor no matter how well they choose. Once we factor in this disparity in initial endowments, we see that it is one-boxers who act irrationally. Endowed with terrific options ($1,000,000 vs $1,001,000), they choose the worst, whereas two-box types respond to their paltry options ($0 vs $1000) by choosing the best. Moral: people who choose irrationally from desirable options can end up better off than people who choose rationally from undesirable options. However, this does not apply in SF, Wells and Spencer argue, because "like consistent left-handers, consistent right-handers always make their choices [in circumstances that involve] exactly [1.05 utiles]." The point seems to be that righties have *better* options in SF: righties cannot do worse than 0.05 utiles, while lefties can end up with 0; righties can secure as much as 1.05, while lefties max out at 1. So, it *should* count against righties that they end up worse off than lefties.

This reasoning is flawed. Spencer and Wells misidentify the advantages of righty versus lefty decisions, and their claim about "consistent" right- and left-handers is only plausible for choices among *types*. For definiteness, suppose Omega correctly predicts lefty choices at a rate 0.2 lower than righty choices, and that he is correct about righties who go to Aleppo/Damascus at a rate of $p_R = 0.9/q_R = 0.7$. Making these assumptions in a real Newcomb Problem means agreeing that lefty-types are 0.2 less likely to die than righty-types *whether they point with their left or their right*, a *huge* initial advantage for lefty-*types*! They enjoy a 0.2 higher survival rate even if they take cake, and righty-types face a 0.2 lower survival rate even if they refuse it. Explicitly, a lefty-type faces one of these decisions, though they know not which (being unsure of their type):

Alep^T & Left^T	Dam^T & Left^T
Alep$_R$ → cake, 0.7 death.	Alep$_R$ → cake, 0.5 death.
Alep$_L$ → no cake, 0.7 death.	Alep$_L$ → no cake, 0.5 death.
Dam$_R$ → cake, 0.3 death.	Dam$_R$ → cake, 0.5 death.
Dam$_L$ → no cake, 0.3 death.	Dam$_L$ → no cake, 0.5 death.

A righty-type faces one of these decisions:

Alep^T & Right^T	Dam^T & Right^T
Alep$_R$ → cake, 0.9 death.	Alep$_R$ → cake, 0.3 death.
Alep$_L$ → no cake, 0.9 death.	Alep$_L$ → no cake, 0.3 death.
Dam$_R$ → cake, 0.1 death.	Dam$_R$ → cake, 0.7 death.
Dam$_L$ → no cake, 0.1 death.	Dam$_L$ → no cake, 0.7 death.

Clearly, it would be better to face one of the top two choices than one of the bottom two, but that bird will have flown by the time you choose. Whatever decision you face, you should take cake since doing so has no effect on your survival probabilities. It only affects what you know about them.

We can, of course, imagine pseudo-Newcombs wherein you should refuse cake because hand-choice causally affects survival, perhaps because you choose your hand-type *before* Omega guesses. Here it is relevant (and decisive) that "consistent" right-handers (righty-types) end up poorer than "consistent" left-handers. But, in SF you choose an act, not a type, only *after* Omega guesses. Once he has guessed, all advantages of being a lefty-type evaporate: pointing with your left has no differential effect except to cost you cake.

Varying this theme, it is easy to confuse the claim that you should choose a lefty *act* in DD with the claim that you should choose a lefty *decision*. Suppose you choose in stages. Initially, you (irrevocably) choose a hand to point with, and then you choose between $\{Alep_R, Dam_R\}$ or $\{Alep_L, Dam_L\}$ depending on the hand selected. At the initial stage, you should clearly choose the Lefty decision. Though this means forgoing cake, it more than compensates by offering a 0.2 better survival probability. It does not follow, however, that you should choose $Alep_L$ or Dam_L in SF, where righty-acts are options. In the context of a choice between the Righty versus Lefty decisions, SF's acts are not options, but potential *consequences*, or *acts-in-prospect*, as in section 3. Recall too that, like EDT, CDT assesses acts-in-prospect by their news values. When assessing the Lefty decision, CDT tells you to (i) figure out how likely you are to choose $Alep_L$ and Dam_L later if you choose Lefty now, and (ii) use these probabilities to determine an expected news value for the decision. Similarly for choosing Righty now. This yields

$$U(\text{Lefty}) = prob(\text{Alep}_L) \cdot V(\text{Alep}_L) + prob(\text{Dam}_L) \cdot V(\text{Dam}_L)$$

$$U(\text{Righty}) = prob(\text{Alep}_R) \cdot V(\text{Alep}_R) + prob(\text{Dam}_R) \cdot V(\text{Dam}_R).$$

To find the relevant probabilities, you use equilibrium values for future decisions. With the reliability rates we've been using, $U(\text{Lefty}) = 0.5 > U(\text{Righty}) = 0.233$.[12] So, unless cake is better than a 0.267 increase in your mortality rate, you should choose to make the lefty decision. Starting with

---

[12] $prob(\text{Alep}_R) = 1/3$, $prob(\text{Alep}_L) = 0$, $V(\text{Alep}_L) = 0.3$, $V(\text{Dam}_L) = 0.5$, $V(\text{Alep}_R) = 0.1$ and $V(\text{Dam}_R) = 0.3$.

different p and q values will yield different utilities, but even for small differences in Omega's predictive abilities, the cake must be terrific to make the Righty decision a rational choice. But, to reemphasize the key point, this does not imply that you should point with your left in SF. In SF all four acts – $Alep_R$, Dam_R, $Alep_L$, Dam_L – are options, and you are assessing them on the basis of their propensity to cause desirable consequences. As long as both righty acts are available, one will always win this competition. Pointing with your left arm when you can use your right is forgoing cake needlessly.

Finally, let's consider a more menacing counterexample, due to Ahmed (2014b).[13] Imagine DD with a perfect predictor and the added option, Coin, of letting your destination be settled by a fair coin (Aleppo iff heads). Omega can predict whether you toss the coin but not how it lands. If he predicted a toss, he rolled a die and sent assassins to Aleppo/Damascus if it landed even/odd. Wherever the assassins are, and whatever sent them there, your chance of avoiding death by tossing is 0.5. So, your decision looks like this:

Table 6

DDC	$Alep^{\pi\pi}$	$Dam^{\pi\pi}$	$Coin^{\pi\pi}$ Head	$Coin^{\pi\pi}$ Tail
Alep	0, 1	1, 0	0, 0	1, 0
Dam	1, 0	0, 1	1, 0	0, 0
Coin	0.5, 0	0.5, 0	0.5, ½	0.5, ½

All equilibria satisfy $0 \leq prob^*(\text{Alep}) = prob^*(\text{Dam}) \leq 0.5$, and $U^*(\text{Alep}) = U^*(\text{Dam}) = U^*(\text{Coin}) = 0.5$. So, CDT requires *indifference* between all acts, which means that you should refuse to pay a cent to toss the coin. In contrast, EDT has you pay any price up to 0.5 since $V(A) = V(D) = 0 < V(\text{Coin}^{-\Delta}) = 0.5 - \Delta$, where $\text{Coin}^{-\Delta}$ is the option of paying Δ utiles to toss the coin.

Ahmed calls this advice "absurd" and sees it is a reason to abandon CDT. (2014: 592). Many will agree. But, CDT has it right: you should not pay a cent to toss the coin because you should not see yourself as buying anything with your money. Proponents of EDT, of course, argue that you are buying an increased probability of life. Since Omega is 100% reliable, they argue, you are

---

[13] Spencer and Wells offer an example, *The Frustrater*, which is equivalent to Ahmed's.

certain to die in Aleppo/Damascus if you choose Aleppo/Damascus. But, if you toss the coin then, wherever Omega's henchmen are, you have a 0.5 objective chance of living. It boils down to a certainty of death versus a 50% chance at life (for a pittance). Pay!

To see the flaw in this reasoning, note that in any equilibrium $prob^*(\text{Alep}) = prob^*(\text{Dam}) = x > 0$ you do *not* believe that Alep and Dam offer certain death. You estimate that picking Alep gives you probability x of death (Alep$^\pi$), probability x of life (Dam$^\pi$), and probability $1 - 2x$ of a fifty-fifty objective chance at life (Coin$^\pi$). Thus, your *credences* for the causal hypotheses H_A = "Choosing Alep will cause my death" and H_D = "Choosing Dam will cause my death" are both 0.5, not 1.0! What trips people up is that, in virtue of Omega's reliability, you *are* justifiably certain that:

> H If I choose Alep or Dam, then choosing the act I choose will cause my death.

H seems equivalent to the conjunction H_A & H_B, and seems to entail that choosing Alep or Dam will cause sure death. This is wrong. Since the phrase "the act I choose" *rigidly* denotes what you *actually* choose, H says nothing about the act not chosen. In fact, choosing the other act will cause your *survival*, which would make it a better choice than Coin$^{-\Delta}$ if you knew what it was. Unfortunately, unlike Omega, you will not know what "the act I choose" and "the other act" denote until after you pick. When you pick, you are constrained by your *current* evidence, on which you assign probabilities of x, x, and $1 - 2x$ to the hypotheses that choosing Alep/Dam will cause your objective chance of death to be 100%, 0%, or 50%, respectively. So, your credence of living conditional on any of these acts is 0.5. In terms of your subjective estimates of survival probabilities, all three acts offer the same thing. So, paying to toss the coin would be paying for what you already take yourself to have.

The idea that you get something for your money has at least two possible sources. It may express an irrational form of *ambiguity aversion*, or it may be a conflation between DDC and two subtly different pseudo-Newcombs in which Coin$^{-\Delta}$ is optimal. First, consider ambiguity aversion, our well-documented preference for credences based on known objective chances. People might prefer Coin because it ensures a 0.5 *objective* chance of survival, while Alep/Dam's 0.5 survival probability reflects uncertainty about the chances: in equilibrium, DDC is like drawing from an urn with balls marked "100% death," "0% death," "50% death" in proportions of x, x, and $1 - 2x$. Choosing Coin replaces ambiguity with clarity. Though I will not argue it here, I see ambiguity aversion as irrational.[14] But, even if I am wrong, it explains Coin's appeal in a way that is consistent with CDT. The ambiguity-averse agent chooses Coin not to improve expected survival probabilities, but to relieve herself of the anxiety of not knowing objective risks.

Paying to toss the coin also seems right because it is so easy to confuse DDC with pseudo-Newcombs where it is right. We will consider two examples. First, suppose you are slated to face DD with a perfect predictor later, but can now *avoid* that choice by paying Δ and going to Aleppo/Damascus iff a fair coin lands heads/tails. Omega has predicted whether you will take this deal. If he guessed that you would accept, he rolled a fair die and sent assassins to Aleppo/Damascus iff even/odd. Otherwise, he executed his usual DD protocol. CDT might seem to advise against paying because each act in DD has an equilibrium utility of 0.5, while $U^*(\text{Coin}^{-\Delta}) =$ 0.5 – Δ. Not so! CDT has you pay up to a half utile to toss to *cause* the desirable result of avoiding the Alep/Dam choice! Instead of choosing from {Alep, Dam, Coin}, you get to decide between $F = $ [DD later, keep Δ] and $\sim F = $ [avoid DD, pay Δ, 50% risk of death]. Since facing-DD-and-choosing-Alep or facing-DD-and-choosing-Dam are causally downstream of your *current* choice, CDT treats them as acts-in-prospect to be assessed by news value. Of course, it is terrible news that you are slated to go up against a perfectly reliable Omega in DD – a sure harbinger of death – and CDT recognizes this by setting $U^*(F) = prob^*(A)\cdot V(A) + prob^*(D)\cdot V(D) = V(F) = 0$, and $U^*(\sim F) = V(\sim F) = 0.5 - \Delta$. So, like EDT, CDT says you *should* pay up to a half utile to avoid DD.

F vs $\sim F$ is easily conflated with DDC. Even Ahmed seems to run them together. When he supposes that you face DD against a perfect predictor, he writes

> Everyone agrees that yours is an unfortunate situation. You are playing high-stakes hide-and-seek against someone who can predict where you will hide... There is every reason to think you will lose. (2014b: 588)

He then imagines a "third option," $\text{Coin}^{-\Delta}$, and shows that in a choice from {Alep, Dam, $\text{Coin}^{-\Delta}$} CDT rejects $\text{Coin}^{-\Delta}$ for any $\Delta > 0$. This is entirely correct, but Ahmed shifts focus when arguing that this advice is absurd:

> Would you rather be playing hide-and-seek against (a) an uncannily good predictor of your movements or (b) someone who can only randomly guess

at them? [You are being] offered the chance to reduce [Omega] from (a) to (b). Of course you should take the offer. (2014b: 589)

I agree! If you can choose to make a pseudo-Newcomb decision with Omega no better than chance rather than a Newcomb decision where he is perfectly reliable, you should do it! This is why you take ~F over F. By choosing ~F, you take Alep and Dam off the table, thereby forcing Omega to "play on neutral turf" where his predictive powers can do you no harm. With Alep and Dam on the table, he has a significant probability ($2x$) of guessing your choice. But, he can be no better than chance if you tie your destination to the coin toss. CDT says to take that deal! But, that deal is *not* DDC. In DDC, Alep and Dam remain live options right up to the time you irrevocably pick, which forces you to ask how likely it is that choosing them will cause your death. The answer is 0.5, the same as your credence that choosing Coin will cause your death. So, you should pay for Coin when that takes Alep and Dam off the table, but when both remain live options, paying to toss is paying for what you already have. Thus, CDT gets both DDC and the ~F vs F decision right.

The distinction between DDC and similar decisions in which CDT endorses paying turns on a subtle difference in options. DDC has three – Alep, Dam, and $\text{Coin}^{-\Delta}$ – and you can only refrain from choosing $\text{Coin}^{-\Delta}$ by choosing one of Alep or Dam. There is no fourth alternative of deciding not to toss the coin without committing (by choice or pick) to Alep or Dam, i.e., no disjunctive "Alep or Dam" option. As a result, you cannot rationally choose $\text{Coin}^{-\Delta}$ unless your estimate of the survival probability caused by tossing the coin exceeds your estimates of the highest of the survival probabilities caused by Alep and Dam, which never happens. In contrast, if you have an option like ~F that lets you decline to toss *without* picking a city, then you assess $\text{Coin}^{-\Delta}$ relative to the menu $\{\text{Coin}^{-\Delta}, \sim\text{Coin}^{-\Delta}\}$, rather than $\{\text{Coin}^{-\Delta}, \text{Alep}, \text{Dam}\}$. Here, $\sim\text{Coin}^{-\Delta}$ is the option of first declining $\text{Coin}^{-\Delta}$ and only later deciding between Alep and Dam. In any decision with this bipartite structure, CDT will treat $\sim\text{Coin}^{-\Delta}$ & Alep and $\sim\text{Coin}^{-\Delta}$ & Dam as acts-in-prospect, and will endorse $\text{Coin}^{-\Delta}$ over $\sim\text{Coin}^{-\Delta}$.

Some people may assume that $\sim\text{Coin}^{-\Delta}$ is always an option. They might even think that a rational agent can always ensure its availability by making a kind of pre-decision in which the options are choosing *later* from the $\{\text{Coin}^{-\Delta}, \sim\text{Coin}^{-\Delta}\}$ menu or the $\{\text{Coin}^{-\Delta}, \text{Alep}, \text{Dam}\}$ menu. I doubt this, but even if it were true, it would pose no problems for CDT. If $\sim\text{Coin}^{-\Delta}$ is always an option, then Ahmed's counterexample will never arise and CDT will always rightly recommend paying to toss the coin. If agents can pre-decide between

{Coin$^{-\Delta}$, ~Coin$^{-\Delta}$} and {Coin$^{-\Delta}$, Alep, Dam}, then we have another pseudo-Newcomb that is easily confused with DDC. CDT will treat all entries in these menus as acts-in-prospect, and will recommend selecting the first menu, and subsequently choosing Coin$^{-\Delta}$ from it. Either way, Ahmed's example does not undermine CDT. In any version of the problem in which ~Coin$^{-\Delta}$ is an option, CDT recommends paying to toss. In DDC, where Alep and Dam are options but ~Coin$^{-\Delta}$ is not, CDT rightly recommends not paying because paying buys nothing.

One might consider other examples, but readers should have the flavor of CDT's responses. If we focus on *real* Newcomb Problems, which satisfy *NP*$_1$-*NP*$_5$, the sophisticated version of CDT that requires choices to be made in equilibrium gets every case right. When it seems to falter, either the theory is being misapplied, options are being misidentified, or a solution to a pseudo-Newcomb Problem is being passed off as a solution to a genuine Newcomb Problem.

8   "Click!" Bait for Causalists

Huw Price and Yang Liu

1 *Euthyphro* for Causalists

Causality and rational means–end deliberation seem to go hand-in-hand, in normal circumstances. Putting it roughly, A causes B if and only if, other things being equal, it would be rational for an agent who desired B to *do* or *bring about* A, in order to realize B. Much of the interest of Newcomb Problems lies in the fact that they seem (to some!) to provide counterexamples – cases in which rational means–end reasoning goes one way, while causal connectedness goes another. "Evidentialists" argue the case for such counter-examples, while "Causalists" oppose it.

Newcomb Problems aside for the moment, let's think about the nature of the relationship between causality and rational action. Let's ask the *Euthyphro* question. Is it the causal connection between A and B that makes it rational to do A to achieve B? Or does the rationality of the latter somehow constitute or ground the fact that A causes B? Let's call the first option the *objectivist* view, the second the *subjectivist* view.[1]

The standard understanding of Causal Decision Theory (CDT) seems to presuppose objectivism. CDT is offered as a formalized prescriptive theory of rational decision, couched in terms of an agent's beliefs about objective relations of causal dependence. It remains a subjective theory in the sense that it is couched in terms of what the agent *believes*, rather than in terms of how the world actually *is*. But it is objectivist *about causation*, in that the crucial beliefs concern objective causal relations.[2]

In our view, this understanding of CDT is not compulsory. Indeed, it cannot be compulsory so long as subjectivism about causation is itself a live

---

[1] These are not necessarily the happiest terms for the views in question. We use them here because they are the terms in use in the case of chance, to which we want to draw an analogy. Elsewhere (Price 2001, 2007) we have recommended the term "pragmatism" for what we here call "subjectivism."

[2] That is, for our purposes, relations whose nature is to be understood independently of their role in CDT itself.

option. A subjectivist will agree that CDT gets things right *in a descriptive sense*, formalising the relation between causal beliefs and rational deliberation. But since the subjectivist takes this relation to be constitutive of causal belief, CDT is in effect being "read backwards," characterizing causality in terms of rational action.[3]

Among the advantages of this subjectivist version of CDT is that it allows a pleasing reconciliation between CDT and Evidential Decision Theory (EDT) – in effect, a dissolution of Newcomb's Problem. This reconciliation is the end goal we have in mind, but our main concern here will be one step back: we want to present an objection to the orthodox objectivist version of Causalism. The objection turns on a dilemma that confronts Causalism, when presented with Newcomb Problems. So far as we know, this dilemma has not been noted in this form, either by Causalists or their Evidentialist opponents.

One horn of the dilemma is to embrace a subjectivist understanding of the relation between causality and rational deliberation – unacceptable for an orthodox objectivist Causalist, but a welcome option for subjectivists themselves. As we'll see, this option either brings CDT into complete alignment with EDT, in standard Newcomb cases, or explains why EDT takes precedence, in versions of the view that allow some small divergence. (The difference between these versions depends on a terminological choice about the use of "causation," in difficult cases.)

The other horn of the dilemma retains objectivism about causality, but at the cost, as we'll argue, of making it mysterious why causality should be thought to be the arbiter of rational action, in the way that CDT proposes.

It turns out that the structure of the options here is closely parallel to those in the case of chance, in a tradition whose subjectivist roots trace back (for our purposes) to the kind of theories of objective chance proposed by D. H. Mellor and David Lewis. (We take our subjective/objective terminology from Lewis.) This analogy between causation and chance has not been widely noticed,[4] perhaps because objectivist intuitions are even more prevalent for causation than for chance, so that it has been hard to see the pressure for a degree of subjectivism that flows from decision theory (a pressure to which both Mellor and Lewis were sensitive in the case of chance).

---

[3] As we shall see in section 4 below, this is compatible with the view that CDT functions prescriptively in normal cases, where the relevant causal relations are not in doubt.

[4] Lewis himself missed it, apparently – a fact reflected, as we'll explain, in his unwavering commitment to two-boxing in the classic Newcomb Problem.

To exhibit this pressure in the causal case, we'll begin with the observation that Newcomb Problems raise the possibility that the facts of causation are not what they seem (in particular, not what the orthodox Causalist assumes them to be), and that this explains the appearance that CDT and EDT offer different recommendations. The idea that Newcomb Problems might involve non-standard causal relations certainly isn't new. (Like so much else in this field, it goes back to Nozick's original (1969) paper, and is explored in more detail in Mackie (1977).) What is new, so far as we know, is a full appreciation of how this possibility turns into a dilemma for orthodox objectivist Causalism, once the *Euthyphro* issue is in view. We'll approach the issue via the work of Michael Dummett, who gets close to the points we want to make.

2 Bringing about the Past?

Dummett published two famous early papers about the direction of causation, ten years apart (Dummett 1954, 1964). Dummett has a similar cluster of concerns in both papers: why we take causes to *precede* their effects in time (or are at least to be *no later than* their effects, as in cases of simultaneous causation); why we think it makes sense to act for the sake of *later* ends but not for *earlier* ends; and, especially, whether the latter principle might coherently admit exceptions – cases in which it would make sense to act to *bring about* an earlier end.

In the 1954 paper Dummett satisfies himself that causation itself is unidirectional. He argues that causes are the *beginnings* of explanatory chains or processes. Their effects are later events in these chains, and hence lie later in time.[5]

> The temporal direction of causation, from earlier to later, comes in because we regard a cause as *starting off* a process: that is to say, the fact that at any one moment the process is going on is sufficiently explained if we can explain what began it. Causes are simultaneous with their immediate effects, but precede their remote effects. (1954: 29)

---

[5] Dummett thinks that the continuation of such processes does not require explanation, once they get going. He doesn't raise the question whether the past–future asymmetry of the picture is merely a conventional matter, on a par with Hume's stipulation that causes are the *earlier* and effects the *later* of pairs of constantly conjoined events. Another possibility would be to tie it to some fundamental directionality in time, but this would sit oddly with his determination *not* to take past–future asymmetries for granted.

Dummett then asks whether, "given that *in general* causality works in the earlier-to-later direction, we could not recognize a few exceptions to this general rule."

> If we find certain phenomena which can apparently be explained only by reference to later events, can we not admit that in these few cases we have events whose causes are subsequent to them in time?

"I think it is clear that we cannot," Dummett replies:

> One event is causally connected with another if it is either its immediate cause or it is one of its remote causes. To be its immediate cause, it must be simultaneous with it. If it is a remote cause of it, then it is so in virtue of being the immediate cause of the beginning of some process whose con-tinuance is not regarded as requiring explanation, and whose arrival at a certain stage is in turn the immediate cause of the event in question. . . . An event subsequent to the event whose occurrence we were wishing to explain could fall into neither of these two categories. (1954: 31)

However, Dummett continues,

> [t]his explanation why an effect cannot precede its cause does not . . . end the matter. We may observe that the occurrence of an event of a certain kind is a sufficient condition for the previous occurrence of an event of another kind; and, having observed this, we might, under certain condi-tions, offer the occurrence of the later event, not indeed as a causal, but as a *quasi-causal* explanation of the occurrence of the earlier.[6] (1954: 31–32, our emphasis)

Dummett then discusses the possibility of such future-to-past "quasi-causation" at considerable length. He imagines the following objection:

> [T]o suppose that the occurrence of an event could ever be explained by reference to a subsequent event involves that it might also be reasonable to bring about an event in order that a *past* event should have occurred, an event previous to the action. To attempt to do this would plainly be nonsensical, and hence the idea of explaining an event by reference to a later event is nonsensical in its turn. (34–35)

---

[6] It is the lack of a suitable embedding in a (past-to-future) process that prevents it from being *genuinely* causal, in Dummett's view.

In response, Dummett argues that the idea of acting to bring about a past event is far from nonsensical. He concludes that in exceptional circumstances it would be not merely coherent but entirely rational to act in this way. The 1954 paper closes with this example of the kind of case he has in mind.

> Imagine that I find that if I utter the word "Click!" before opening an envelope, that envelope never turns out to contain a bill; having made this discovery, I keep up the practice for several months, and upon investigation can unearth no ordinary reason for my having received no bill during that period. It would then not be irrational for me to utter the word "Click!" before opening an envelope in order that the letter should not be a bill; it would be superstitious in no stronger sense than that in which belief in causal laws is superstitious. Someone might argue: Either the envelope already contains a bill, or it does not; your uttering the word "Click!" is therefore either redundant or fruitless. I am not, however, necessarily asserting that my uttering the word "Click!" changes a bill into a letter from a friend; I am asserting (let us suppose) that it prevents anyone from sending me a bill the previous day. Admittedly in this case it follows from my saying "Click!" that if I had looked at the letter before I said it, it would not have been a bill; but from this it does not follow that the chances of its being a bill are the same whether I say "Click!" or not. If I observe that saying "Click!" appears to be a sufficient condition for its not being a bill, then my saying "Click!" is good evidence for its not being a bill; and if it is asked what is the point of collecting evidence for what can be found out for certain, with little trouble, the answer is that this evidence is not merely collected but brought about. Nothing can alter the fact that if one were really to have strong grounds for believing in such a regularity as this, and no alternative (causal) explanation for it, then it *could not but be rational* to believe in it and to make use of it. (1954: 43–44, emphasis added)

2.1 From Dummett to Newcomb

Dummett's "Click!" example is easily transformed into a Newcomb Problem. We simply need to suppose that Dummett incurs a small cos, say tuppence, for saying "Click!". Not saying "Click!" then *dominates* saying "Click!" – whether or not the envelope contains a bill, Dummett does better by saving his tuppence. Presumably Dummett will still feel that "it *could not but be rational*" to say "Click!" despite the expense. (We are assuming that the usual expense of Dummett's bills, averaged over all his incoming mail, was more than tuppence per envelope – even in 1954.)

If we construe the example in this way, Dummett counts in modern terms as an Evidentialist. What will he say to someone who disagrees about the rationality of paying to say "Click!"? With some violence to his own Oxford idiolect, we can imagine him offering the classic Evidentialist response: "If you're so smart, why ain'cha rich?" And such an opponent – a Causalist, in our terms – will offer the standard reply. In each individual case, Dummett could have saved his tuppence. As Dummett himself says, "it follows from my saying 'Click!' that if I had looked at the letter before I said it, it would not have been a bill" (1954: 44). So, claims the Causalist, Dummett always chooses the sub-optimal option. *Contra* Dummett, it could not but be *irrational* to waste one's money on the "Click!".

Causalists shouldn't relax, however. As we noted, Dummett himself calls what the "Click!" case displays "quasi-causation." By his later piece he's dropped the "quasi," and talks explicitly about "backwards causality":

When, however, we consider ourselves as agents, and consider causal laws governing events in which we can intervene, the notion of backwards causality seems to generate absurdities. If an event *C* is considered as the cause of a preceding event *D,* then it would be open to us to bring about *C* in order that the event *D* should have occurred. But the conception of doing something in order that something else should have happened appears to be intrinsically absurd: it apparently follows that backwards causation must also be absurd in any realm in which we can operate as agents. (1964: 340)

The rest of the 1964 paper is devoted to rebutting this appearance of intrinsic absurdity, covering similar ground to the earlier piece. Dummett does not re-use the "Click!" example, but he does offer a more complicated example with essentially the same structure.

But if we allow this shift in terminology (from "quasi-causation" to "causation" simpliciter), then Causalism will recommend one-boxing, apparently, in the same case. Dummett imagines that by saying "Click!" he *brings it about* that the envelope does not contain a bill. The 1964 paper is called "Bringing about the past," and Dummett's main claim is that if there were such an exceptional case, we would be entitled to regard ourselves as bringing about the past in precisely the same sense as, ordinarily, we regard ourselves as bringing about the future.

We can now see why Dummett represents such a challenge to conventional two-boxing Causalism. By putting the causal terminology itself into play in this way, he threatens to defend one-boxing with the Causalist's own weapons. The Causalist can object to the terminological move, of course, but this won't be straightforward. She will be trying to defend causal orthodoxy in a case in which the rationality of an associated action is in dispute. Nancy Cartwright (1979) taught us to associate the distinction between causal and merely associative correlations with the distinction between effective and ineffective strategies. But in the present case, Dummett and the two-boxer Causalist disagree about whether paying to say "Click!" is an effective strategy, so there's no easy way to apply Cartwright's criterion to resolve the terminological issue. (This thought will lead us to our dilemma.)

True, there's still the matter of the counterfactuals. Could they get the Causalist off the hook? We'll come back to this (as Dummett himself did, in later work), but is worth noting that even in the 1954 paper, Dummett notes that our counterfactual intuitions can't be trusted in this kind of case:

> [W]here we are concerned with a regularity which works counter to ordinary causal regularities, our normal methods of deciding the truth of a contra-factual conditional break down. (1954: 38)

2.2 Dummett's Dilemma

Dummett's trajectory between 1954 and 1964 thus highlights a difficulty for orthodox CDT – one that gets far too little attention in discussions of Newcomb Problems, in our view. It turns on the fact that Newcomb Problems are, by their nature, cases in which the relevant causal facts may be disputed. When we unpick the options in such a dispute, we find the dilemma that confronts Causalism.

Why isn't this issue already well known? Perhaps because neither side has seen a motivation to explore it. It is an unwelcome option for the orthodox objectivist Causalist, who has an obvious interest in assuming that the causal facts are sufficiently solid to ground rational decision theory, in difficult cases; and an apparent irrelevance for the Evidentialist, who professes not to be interested in causation in the first place. So, it lies in something of a blind spot, an unappealing corner from both points of view. Yet in that blind spot lies something important for both sides, in our view: the prospect of a reconciliation over Newcomb's Problem that endorses the key insights both of Causalism and Evidentialism.[7]

---

[7] More importantly still, this reconciliation is the clue to an important ingredient of an understanding of causation itself, in our view.

To expose the dilemma for Causalism, imagine the kind of case Dummett points us to, where we have two views of causation on the table. They differ in extension, so cannot both be a true guide to rational decision, by CDT's lights, if "rational" is to remain univocal. How are we to adjudicate? The Causalist wants the understanding of causation that gives the rational choice, when plugged into CDT. But how are we to know which is the rational choice, given that we subscribe to CDT, and we are unclear about the meaning of "cause"? We have an equation with too many variables.

At the heart of this problem lies our dilemma. On one horn, the Causalist attempts to meet the challenge to her use of causal terminology in objectivist terms. In other words, she relies on some principle *other than an explicit rationality criterion* to adjudicate the use of "cause" and related terms. For example, she insists that causes be no later than their effects. To defend CDT, she then faces the problem of explaining why a rule for rational action should always respect this criterion. Dummett's "Click!" case will be a direct challenge.[8]

On the other horn, the Causalist relies on an explicit rationality criterion to fix the extension of causal terminology – in effect, she opts for subjectivism, reading CDT "back to front," as an analytic constraint on the notion of causal dependence. But now CDT is no use at all as a guide to the rational course of action, in disputed cases. On the contrary, a ruling on the application of causal notions must lie *downstream* of the resolution of such disputes, in contested cases. (Again, "Click!" provides an example.) And if the subjectivism itself weren't bad enough, the only available account of rationality will be that of the Evidentialist, presumably.[9] Swept away downstream, Causalists risk being washed ashore in the land of one-boxers.

In summary, this is the dilemma. If the principle criterion of causality is not that of supporting rational action as such, then the issue can be raised as to whether (and if so why) the criterion always keeps step with rational action. If the criterion is rational action, then the causal structure cannot be resolved while rational action is in dispute – and that way lies subjectivism, and arguably Evidentialism.

---

[8] Dummett's method seems to be applicable to whatever criterion the Causalist proposes at this point. We simply imagine a case in which we encounter a reliable correlation that doesn't meet the criterion in question, but which nevertheless seems to be of practical value in the way that saying "Click!" is in Dummett's case.

[9] A subjectivist can't characterise rationality in causal terms, at least initially – that would simply be circular.

In honor of Dummett, we'll call this *Dummett's Dilemma* (DD). Dummett himself does not describe DD explicitly. Indeed, as we'll see in a moment, he makes a move that prevents him from developing its full potential as an objection to Causalism. But he does exemplify it, in the sense that the problem he represents for an orthodox two-boxing Causalist lines up with the first horn in 1954 and the second horn in 1964. In 1954 the implicit challenge is: Why is *Causal* Decision Theory rather than *Quasi-causal* Decision Theory the appropriate guide to rationality in the "Click!" case, if the difference between the two rests merely on a criterion couched in terms of the direction of time? In 1964 it is: On what grounds *independent of a knowledge of causal structure* can we declare the "Click!" policy irrational, in order to deny that there is backward causation?

2.3 Dummett on Counterfactuals

Let's return to the counterfactuals. Do they provide a middle way for object-ivist Causalism, an objective criterion for causal connection that wears its link to rationality on its face? It may seem so. Imagine this objection to Dummett's reading of the "Click!" example. "But Dummett," says the opponent,

> if you had not said "Click!" the envelope would still not have contained a bill, and you would have been tuppence the richer. The truth of that counterfactual shows that there's no causation involved, and explains the sense in which you did the irrational thing!

This is the argument for irrationality involved in the standard Causalist response to the "Why ain'cha rich?" challenge (see Lewis 1981b).

We know how Dummett would have responded to this objection, for he discusses Newcomb Problems in two papers (1986, 1987) from the mid-1980s. He retains his commitment to one-boxing, and his main response to the two-boxer's appeal to counterfactuals is to deny that ordinary counter-factual reasoning is a helpful guide to rationality in such cases.[10]

---

[10] In general, Dummett thinks that counterfactuals are a great deal more obscure than many philosophers tend to assume:

> [T]he meaning of a subjunctive conditional is enormously obscure; they have been a thorn in the side of analytic philosophers for many decades. Even more obscure than their meaning is their point: when we have devised some theory to explain their truth-conditions, we are at a loss to explain for what purpose we want to have sentences with those truth-conditions in the language at all. Very important things may hang on them: whether the accused is guilty of murder may depend on whether the deceased would still have died if the accused had not acted as he did. But

I thus have a choice between doing something that will, with a very high probability, result in my getting $1,000 and doing something that will, with a very high probability, result in my getting $10,000. Plainly, ... the rational thing for me to do is the second. After I have done it, the rules governing the assertion of counterfactual conditionals may entitle me to assert, "If I had taken both boxes, I should have got $11,000"; *but that is only a remark about our use of counterfactual conditionals.* Before I make my choice, I should be a fool to disregard the high probability of the statement "If I take both boxes, I shall get only $1,000." That is not merely a remark about our use of the word "probability," nor even about our use of the word "rational," but about what it is rational to do. (1986: 375, emphasis added)

In our view, however, this leaves Dummett vulnerable to the objection that the ordinary usage of counterfactuals and the ordinary usage of the notion of rationality are part of the same package. For example, it might be held to be an essential criterion for the rationality of an action that one not be in a position to know in advance that if one does it, one would have done better by doing something else. This claim might be disputed, but while it is in play, Dummett will not be free to dismiss the relevance of ordinary counterfactual claims, without also threatening the ordinary use of claims about rationality. In other words, his own claim about "what it is rational to do" will be just as suspect (despite his insistence to the contrary).

At one point, Dummett makes what we take to be the right move against the Causalist appeal to counterfactuals. Describing a slightly different version of the Newcomb Problem, Dummett says this:

Smith [the two-boxer], when his turn comes, argues, "The money is already in the boxes. Whether each contains £10 or £1,000, I shall get twice as much if I choose to open both boxes"; he accordingly does so, and receives £20 in all. Jones [the one-boxer] says to him, "You were a fool: you should have chosen to open only one box", but Smith replies, "If I had opened only one, I should have got no more than £10." (1986: 356)

So far this is the standard exchange, but Dummett now notes that the one-boxer can challenge the two-boxer's counterfactual:

that is no answer to the question about their point: for, in such a case, we have given the subjunctive conditional such a point, and the question is why we should do so. (1986: 355)

Jones may well retort, "Not at all: if you had opened only one, the psychologist would previously have put £1,000 in each."[11] (1993 (1986): 356)

But Dummett doesn't press the advantage home, and slides back to his dismissive line about the ordinary use of counterfactuals:

> Now which [of Smith and Jones] is right? If the question is, which of their judgements about the truth of the counterfactual conditionals best accords with our established ways of judging such matters, it is arguable that Smith has the better case. If so, all that follows is that he has a more accurate grasp of the existing use of such sentences: it does not show that Jones will be acting foolishly when his turn comes, and he chooses to open only one box, obtaining £1,000. (1993 (1986): 356)

As we said, this lays Dummett open to the charge that counterfactuals and rationality are part of the same package, and that he can't dismiss one in this way without undercutting his own right to use the other "full voice."

A better move, in our view, is to press the dilemma for counterfactuals, just as it has been pressed for causation. In other words, it is to point out that the counterfactual characterization of the case, like the causal characterization is simply contested. The objectivist Causalist now faces the same dilemma as before. If she seeks to resolve the contest by appealing to some independent criterion for the relevant kind of counterfactual dependence – e.g., a temporal asymmetry principle – the question will arise, as in Dummett's classic discussions in (1954) and (1964), as to whether and why that criterion should constrain rational deliberation. But the alternative is subjectivism, with the same consequences here as in the causal case. So the appeal to counterfactuals adds nothing new.

---

[11] Causalists may object that Jones is using a "backtracking" counterfactual, and so not really disagreeing with Smith, who had the regular kind of counterfactual in mind. But this misses the point. Jones's counterfactual is indeed backtracking in the strictly temporal sense, but (at least as we are reading Dummett here) the question as to whether it is backtracking in the causal sense is precisely what is at issue. (The term "backtracking" is unhelpfully ambiguous in this context.)

The same point is missed in a classic discussion by Horgan (1981). Horgan points out that there is a kind of conceptual circularity in the Causalist's defense of the rationality of her decision rule, since it always comes back to a counterfactual: "If you'd made the other choice, you would have done better." He argues that Causalists have no non-circular way to justify such an account of rationality, as opposed to one based on the Evidentialist's "backtracking" counterfactuals. He is right, but he misses the stronger option of putting in play the "causal" notion of counterfactual dependence itself – of regarding that notion as contested in Newcomb cases. This option is the one on which the counterfactual version of DD depends.

Dummett himself shows the same skepticism about the relevance of our use of the term "cause" as about that of counterfactuals – in both cases, he thinks that there is a danger of missing the interesting and important point, which is about whether a certain kind of practical action makes sense:

> Particularly is this so if you concentrate on how the word "cause" is used, and how it's connected with a temporal direction. Now, if someone wants to know whether it's reasonable for him to do something when his motive is that something should have happened—which some people regard as absolutely ridiculous—it's a cheat to fob him off with explanations in terms of how we use the word "cause." That's not the question. Whether it would be called a cause, or is rightly called a cause, may be an interesting question, but it is not the question he was asking. If it's not called a cause, then all right, perhaps we'll call it something else. But the question is: Is there any sense in doing this thing? (1993 (1986): 369–370)

Dummett seems just to take it for granted that the latter question ("Is there any sense in doing this?") is to be answered in probabilistic terms. As he puts it in one of the passages above:

> I thus have a choice between doing something that will, with a very high probability, result in my getting $1,000 and doing something that will, with a very high probability, result in my getting $10,000. Plainly, ... the rational thing for me to do is the second.

Here Causalists may feel that Dummett is simply blind to the main motivation for Causalism, namely, the observation that not all probabilistic dependencies between Acts and Outcomes are sound bases for effective strategies. (They may also feel that his formulation here – "something that will, with very high probability, *result in* ..." – simply builds in the causal notions that they themselves strive to make explicit.) So Evidentialists who wish to exploit DD need to make clear that they can steer clear of these objections. We shall now propose a way to do this, and at the same time to take maximum advantage of DD. The key is to allow the Causalists themselves to do the heavy lifting.

3 Dummettian Evidentialism

Let us introduce a character we'll call the Dummettian Evidentialist, or DEVI, for short. DEVI's strategy for dealing with Causalists goes like this. When offered something claimed to be a Newcomb Problem – i.e., a case in which Evidentialism and Causalism are claimed to recommend different courses of action – she initially plays dumb, claiming to prefer the one-box option (or its analog). Preparing to try to trap the Causalist in Dummett's Dilemma, she looks ahead to a hypothetical series of cases and says, "Why won'cha get rich?"

In some cases, Causalists will be able to explain why DEVI wouldn't get rich (or richer than they themselves would), after all. In the Smoking Gene case, for example, the correlation between smoking and cancer is taken to be explained by a gene that predisposes one to smoking and cancer – it is a common cause of both. This means that someone who would otherwise choose to smoke, but who gives up smoking "to lower their probability of getting cancer," as they put it, will have the gene (and hence get cancer) just as often as any smoker. In this case, their reason for giving up smoking screens off the correlation between smoking and the gene. So, no advantage accrues to such a person, in this kind of case, to offset the disadvantage of losing the pleasure of smoking. Giving up smoking is an *ineffective* strategy for avoiding cancer, in this case.

In this example, choosing not to smoke is the analog of one-boxing – in other words, it is the option favored (or said by Causalists to be favored)[12] by Evidentialists. Unlike in the classic Newcomb Problem, however, there is simply no analog of the "Why ain'cha rich?" challenge for the Causalist to answer. It is not that folk who give up smoking "to lower their probability of getting cancer" do better than Causalists in having lower rates of cancer, though worse than they themselves might have done if they had not needlessly deprived themselves of the pleasure of smoking. They simply do *worse* than Causalists, full stop. They have the same rate of cancer, and less pleasure from smoking.

In a case of this kind, then, the Causalist is able to argue that there is no gap between the two horns of Dummett's Dilemma. Even a subjectivist will have to accept that choosing not to smoke does not have a causal influence on rates of cancer. To argue this way, the Causalist needs to be able to establish that one-boxing (i.e., in this case, refraining from smoking) is an ineffective strategy, *without appealing to* (what she takes to be) *the known facts about the causal structure of the situation*. Why? Because otherwise it will remain an

---

[12] The qualification is important. On the Evidentialist side a long tradition is devoted to arguing that EDT is not committed to one-boxing (or its analog) in many claimed Newcomb cases, including so-called medical Newcomb Problems such as Smoking Gene. The well-known Tickle Defense is one argument of this sort. The beauty of DEVI's strategy is that she makes the Causalist do all the work.

option for DEVI to argue that the Causalist is simply wrong about the causal facts, as below.[13]

In cases in which Causalists are able to respond in this way to "Why ain'cha rich?" – i.e., in which they can reliably predict that DEVI will not get rich, *without relying on disputed causal claims* – DEVI helps herself to the explanation, modifying her Evidentialism, if necessary. In this way, DEVI allows Causalists to do the kind of work that has been done by her own less lazy Evidentialist cousins, of explaining why EDT doesn't in fact recommend one-boxing, in some (claimed) Newcomb Problems, such as Smoking Gene.

A second class of cases are those Newcomb Problems in which the Causalist needs another response to "Why ain'cha rich?" – cases in which both sides agree that the Evidentialist does better in the long run, but in which the Causalist maintains that the Evidentialist *would have* done even better, had she two-boxed. In this case DEVI plays Dummett's card.[14] She says: Here is an alternative view of the causal and counterfactual facts, according to which one-boxing is rational, by your Causalist lights. (This is the same as Dummett saying that saying "Click!" *causes* rather than merely *quasi-causes* the fact that the envelope does not contain a bill.)

The Causalists now face Dummett's Dilemma. If they opt for objectivism, relying on some criterion other than the rationality of two-boxing as the basis for their view of the causal facts, they meet the analog of Dummett's question about the criterion that causation only works "forwards": Why should rational action be so constrained? Whereas if they opt instead for subjectivism, then they are unable to dismiss DEVI's alternative construal of the causal facts, while the issue of the rationality of one-boxing remains in dispute.

DEVI always has the upper hand here, because the riches are on her side. The standard Causalist response to "Why ain'cha rich?" won't work, so long as the counterfactuals are just as much in play as the facts about causation. "On my understanding of causation, and hence counterfactuals," DEVI insists,

---

[13] More generally, when Cartwright argues for claims of this form:

> [C]ausal laws cannot be done away with, for they are needed to ground the distinction between effective strategies and ineffective ones. . . . [T]he difference between the two depends on the causal laws of our universe, and on nothing weaker. (1979: 420)

she requires that the fact that a strategy is effective or ineffective be *prior* to the causal facts that she takes to explain why it is one rather than the other. Otherwise she would have no distinction between *explanans* and *explanandum*.

[14] Or, to put it more accurately, a card we found in Dummett's hand. We saw that Dummett himself does not play the hand to best advantage.

"It simply isn't true that had I two-boxed I would have been richer – on the contrary, I would have been poorer."

For Causalists it is no use trying to stipulate their way out of the problem, by saying in their description of the Newcomb Problem, "We assume that causation works in the ordinary way." For DEVI will respond: "You're not entitled to assume that causation works in the normal way, in proposing examples that put this kind of pressure on the ordinary notion of causation. My point is that so long as you concede that the one-boxer will get rich, you've conceded that the case is in that contentious territory. At that point, my move is on the table, and you can't avoid it."[15]

True, there may well be cases in which it is simply unclear who gets rich. Perhaps there are limits on the long run, or confounding correlations to take into account, or perhaps the example is simply under-described or incoherent in some crucial respect. No matter, from DEVI's point of view. So long as the cases remain unclear or incoherent, they don't constitute a counterexample to the general method; and whichever way they fall, DEVI's method will accommodate them.

4 The Analogy with Chance

We began with the *Euthyphro* question about causation. Is it the causal connection between *A* and *B* that makes it rational to do *A* to achieve *B*? Or does the rationality of the latter somehow constitute or ground the fact that *A* causes *B*? There is an analogous question about chance. Is it the fact that there is a high chance that *P* that makes it rational to hold a high credence that *P*? Or is the fact that there is a high chance that *P* somehow *constituted by* or *grounded in* the fact that it is rational to be confident that *P*?[16] Again, we might call these options *objectivism* and *subjectivism* about chance, respectively.

In the case of chance, one writer sensitive to these issues is D. H. Mellor. Mellor (1971) defends a version of what we are here calling the subjectivist option, calling it "personalism". Following Kneale (1949), he insists that

---

[15] The Causalist can certainly stipulate that the case involves no unusual causation *in her preferred objectivist sense*. But she thereby places herself on the objectivist horn of DD, facing the challenge of explaining why causation in *that objectivist sense* should be a good guide to rational action in an unusual case of the kind presented here (a case in which both sides agree that it is the Evidentialist who gets richer).

[16] It may seem that in the case of causation we are asking about a connection with rational *action*, whereas in the case of chance we are asking about a connection with rational partial *belief*. As we shall see in a moment, however, this difference turns out to be superficial.

personalism is compatible with the view that chances are real and objective – it is just that in saying what they are, we need to begin with rational degrees of belief, or credences.

> [C]an we not analyse full belief that the chance of heads on a coin toss is 1/2 without reference to some supposedly corresponding partial belief that the coin will land heads? The reason for denying this is the fact to which Kneale himself draws attention (p. 18) 'that knowledge of probability relations is important chiefly for its bearing on action'. It follows as Kneale says (p. 20) that 'no analysis of the probability relation can be accepted as adequate . . . unless it enables us to understand why it is rational to take as a basis for action a proposition which stands in that relation to the evidence at our disposal'. Similarly with chance. It must follow from our account that the greater the known chance of an event the more reasonable it is to act as if it will occur. This concept of a quantitative tendency to action is just that of partial belief [i.e., credence] as it has been developed by the personalists. It is thus available to provide in our account of chance that necessary connection with action on which Kneale rightly insists. *A great difficulty facing other objective accounts of chance, notably the frequency theories, has been to build such a connection subsequently on their entirely impersonal foundations.* (Mellor 1971: 3, emphasis added)

Lewis (1980) also defends a form of subjectivism. Like Mellor, he takes chance to be real and objective, but takes it to be definitive of chance that it plays a distinctive role in guiding credence. As he says, he is "led to wonder whether anyone *but* a subjectivist is in a position to understand objective chance!" (1980: 84). Returning to this theme in later work, he criticises rival approaches on the grounds that they pay insufficient attention to this connection between chance and credence: "Don't call any alleged feature of reality 'chance' unless you've already shown that you have something, knowledge of which could constrain rational credence," he says.[17]

---

[17] It is worth quoting the full passage in which Lewis makes this remark:

> Be my guest—posit all the primitive unHumean whatnots you like. . . . But play fair in naming your whatnots. Don't call any alleged feature of reality 'chance' unless you've already shown that you have something, knowledge of which could constrain rational credence. I think I see, dimly but well enough, how knowledge of frequencies and symmetries and best systems could constrain rational credence. I don't begin to see, for instance, how knowledge that two universals stand in a certain special relation N* could constrain rational credence about the future co-instantiation of those universals. (Lewis 1994b: 484)

Lewis gives a name to the principle that knowledge of chance properly constrains rational credence. In view of its centrality to an understanding of chance – "this principle seems to me to capture all we know about chance," as he puts it – he calls it the Principal Principle (PP). His original formulation of PP is as follows:

> Let C be any reasonable initial credence function. Let t be any time. Let x be any real number in the unit interval. Let X be the proposition that the chance, at time t, of A's holding equals x. Let E be any proposition compatible with X that is admissible at time t. Then $C(A|XE) = x$. (1986 [1980]: 87)

(The notion that Lewis here calls the "admissibility" of the evidential proposition E will play a crucial role in what follows, and we'll return to it shortly.)

PP may seem at first sight to be a psychological principle, but this is misleading. Credence is here taken to be *defined* in behavioral terms – "a quantitative tendency to action," as Mellor puts it.[18] So PP is actually a principle describing the rational bearing *on action* of beliefs about objective chance.

With this reminder in place, it is possible to see how, as Price (2012) proposes, there is an analogy between PP and CDT. PP tells us how a rational agent will act in a certain range of decision problems, in the light of her beliefs about objective chance. CDT tells us how a rational agent will act in a different range of decision problems,[19] in the light of her beliefs about objective causal dependency. In both cases we have a principle that connects a belief about what we might call a modal fact, on the one hand, to rational action, on the other.

The analogy between PP and CDT has been obscured by two factors, in our view. Both are products of the particular history of the relevant part of decision theory, and especially of the long rivalry between CDT and EDT. The first factor is that there is no analog, agreed on all sides, of the

---

Lewis's point here is closely analogous to the objectivist horn of DD. A Causalist who rejects subjectivism needs some other criterion for causal connectedness. Following Dummett's lead in the case in which the criterion invokes temporal direction, DD then asks the Causalist to explain why *that* criterion should constrain rational action. This is the same challenge that Lewis here offers to his own objectivist opponents.

[18] Similarly, Lewis stresses that credences simply *are* behavioral dispositions: "No wonder your credence function tends to guide your life. If its doing so did not accord to some considerable extent with your dispositions to act, then it would not be your credence function. You would have some other credence function, or none" (1980: 109).

[19] In fact, a *broader* range.

behaviorally defined notion of credence. In conditional decision theory, where an agent needs to consider the possibility that States might depend on Acts, we are usually offered instead two *different* accounts of the relevant notion of conditional dependence: an evidential one associated with EDT and a causal one associated with CDT. Comparing this to the case of chance, it is as though the behaviorally defined notion of credence had dropped out of unconditional decision theory, and we were being offered two rival accounts of how beliefs about something else constrained action: PP, in which the "something else" is objective chance, and some rival principle, featuring some more evidential rival to chance.

This first factor is easily remedied. We simply need to introduce into our notation a notion of conditional dependence that is understood in the required behavioral, descriptive sense – explicitly neutral between CDT and EDT, who can now disagree about how these behaviorally defined conditional credences (BCCs) are rationally constrained by beliefs about something else.

In the unconditional case, Savage's Decision Theory (SDT) does the job of formalizing a definition of credence in terms of choice behavior. Accordingly, we can then think of PP as a sum of two components: the first a strictly psychological principle (call this PP_ψ) relating beliefs about chances to credences; and the second SDT, connecting credence to choice behavior. We thus have

$$PP = PP_\psi + SDT.$$

To make the analogous move in the conditional case, we need to think of CDT as a sum of two components. The first (call it CDT_ψ) describes how a rational agent aligns her BCCs with her beliefs about relevant causal dependencies. The second is a decision theory that does the job done by SDT in the unconditional case of representing BCCs in term of decision behavior. Fishburn (1964) developed a conditional version of SDT which is a generalisation of Savage's system, and this appears to be what we need. Calling it FDT, this gives us the following analog of the decomposition of PP:

$$CDT = CDT_\psi + FDT.$$

We said earlier that there are two factors that seem to have obscured the analogy between CDT and PP. The second factor is the comparative rarity, within the philosophy of causation, of the kind of subjectivism represented in the case of chance by major figures such as Mellor and Lewis; perhaps combined with the view, from the CDT side in the CDT/EDT debate, and relatedly from Cartwright, that objectivism about causation does useful work in distinguishing effective from ineffective strategies. This factor (or combination of factors) is harder to remedy, but we have offered DD and the analogy with chance by way of therapy. Combined with the salutary analog of Mellor's and Lewis's insistence that unless an account of chance puts the link with action in at the beginning, it will have trouble getting it out at the end, DD leaves the objectivist Causalist in an unenviable position: unable to explain why a decision-maker should *care* about relations of causal dependence, as the Causalist conceives them (at least in the disputed cases in question).

This therapy is combined with the offer of what we hope will prove a palatable alternative for the Causalist, at least with the analogy with chance in mind. Provided we approach causation in a subjectivist spirit, we can enshrine CDT as the, or at least *a*, central commitment of a theory of causation, as Lewis does for PP in the case of chance. Causal relations become objectified relations of conditional credal dependence, the latter understood from the agent's point of view.

Note that this allows CDT a prescriptive and explanatory role, in normal circumstances. In such circumstances we can read off our rational BCCs from our knowledge of causal relations, just as PP enables us to read off our rational (unconditional) credences from knowledge of chances.

What about abnormal circumstances? Here, too, the analogy with chance is helpful. Lewis takes PP to admit exceptions: unusual kinds of evidence can give an agent reason to hold credences that do not align with her beliefs about chance. Lewis calls this "inadmissible evidence"[20] – the evidence that a crystal ball would provide, say, if it showed us the result of a coin toss in advance. The coin might be known to be fair, but we would be foolish to hold a credence of 0.5 in Heads and bet accordingly if the crystal ball reveals that it will in fact land Tails. Lewis thus allows exceptions to PP. But as Hall (1994, 2004) points out, there is another alternative. For Hall, cases of (so-called) inadmissible evidence are just cases in which the chances are strange. PP holds with full generality.

Hall and Lewis agree about the rationality of taking the crystal ball into account. They both disagree with a hardliner who neither modifies her view of chance, nor admits exceptions to PP, in the light of inadmissible evidence. "To be sure," this hardliner says, "my crystal ball shows me that this fair coin

---

[20] An infelicitous name, perhaps, since Lewis is not denying that we should admit it and use it. It is inadmissible only as a constraint on chance, in Lewis's view.

will land Tails on the next toss, and I have no reason to doubt it. But it is a fair coin, so the rational credence in Heads is 0.5. I'll accept an evens bet on Heads."[21]

Note that in disagreeing with the hardliner, Lewis and Hall cannot be relying on PP, for in the cases in question they disagree both about PP's applicability and about the appropriate theory of chance. They therefore share some conception of practical rationality that is for both of them "prior" to the theory of chance. Again, this is the analog of the point on which DEVI's strategy relies, that even an objectivist needs a non-causal criterion for the distinction between effective and ineffective strategies.

For a subjectivist about causation the same options exist with respect to CDT in Newcomb cases. One option (Dummett's choice in 1954) is to allow that like PP for Lewis, CDT is not exceptionless. In strange cases such as "Click!" and the classic Newcomb puzzle, it may be rational to ignore CDT. The other option (Dummett's choice in 1964) is to say as Hall says for chance that causation may be weird in these strange cases, so that CDT recommends one-boxing (and paying tuppence to say "Click!"). Note that just as Lewis and Hall agree that it is rational to let crystal balls influence one's credences, these two options agree that one-boxing is the rational choice in the classic Newcomb Problem. The disagreement is entirely about a terminological matter, the very matter on which Dummett changed his mind between 1954 and 1964.

The peculiarity of the causal case is that the hardline position – unoccupied, so far as we know, in the case of chance – remains the orthodoxy for causation. Such is the grip of objectivism in this case. Against these hardliners we have offered the pragmatic subjectivist wisdom of Dummett in the causal case, and that of Mellor, Lewis and others in what we have urged is a closely analogous case, that of chance. We conclude that objectivist Causalism is philosophically as well as financially impoverished, and that subjectivism takes home the bacon, in both respects.[22]

---

[21] Confronted with the "Why ain'cha rich?" objection, the hardliner makes the same response as the two-boxer in the classic Newcomb Problem: "The reason why [I am] not rich is that the riches were reserved for the irrational" (Lewis 1981b: 377).

[22] We are grateful for comments from an audience at the Munich Center for Mathematical Philosophy.

9　Game Theory and Decision Theory (Causal and Evidential)

Robert Stalnaker

The debates about the proper rational response to the Newcomb Problem have focused on two contrasting versions of Bayesian decision theory, but there is a different formal theory of decision-making – game theory – that developed somewhat independently. This chapter is about that framework. It considers whether it might throw some light on the debates about the Newcomb Problem, and more generally about the role of causal structure in the characterization of rational decision and the relative merits of the causal and evidential formulations of individual decision theory.

I will begin with an exposition of the basic ideas of game theory, looking (in section 1) at the classic development of the theory in the 1940's and 1950's, and then (in section 2) at the later understanding of game theory as an application of Bayesian decision theory. I will then consider, from the perspective of game theory, some of the examples and arguments in the debates about Causal vs. Evidential Decision Theory: in section 3, I will look at Newcomb's Problem itself, and the analogy with one of the most famous games, the Prisoners' dilemma; in section 4, I will consider decision problems that are paradoxical from the perspective of causal decision theory; section 5 will discuss some examples of sequences of decisions, and the way rational agents should think, in their deliberations, about future decision problems they may face as a result of the immediate choices they are deliberating about.

1 Classical Game Theory

A game is a sequence of interacting decision problems, normally, but not necessarily, with more than one agent. The problems interact in that the ultimate outcome (the payoffs for the different players) will normally depend jointly on the actions of the different players. A formal theory of such sequences was developed by John von Neumann and Oskar Morgenstern.[1]

---

[1]　Von Neumann and Morgenstern 1944.

In their theory, the basic representation of a game (a game in *extensive form*) is given by specifying a (finite) set of players, and a finite tree structure, where each node of the tree represents a choice point for one of the players, and the branches from that node represent the alternative available actions for the player. The non-terminal nodes of the tree are labeled by the name of the player who has the choice at that point, and the terminal nodes (representing the possible end points of the game) are labeled with a sequence of utility values, one value for each player, representing the payoffs that motivate the players' choices throughout the game.

Some games will have a chance element, and this is represented by adding a dummy player ("Nature") whose moves are determined by a probability assignment to the branches from any node labeled by that dummy player.

So far, what we have defined is a *perfect information game*: in a game of this kind, each player knows at each of her choice points where she is in the tree, and so what moves other players have made leading up to that choice point. But we also want to represent games in which players are in some cases ignorant of one or more previous moves by other players. For this purpose we add to the formal representation of the game the possibility that different nodes controlled by a single player may be grouped together in *information sets*, representing the fact that the player does not know which of the nodes in the set he is choosing from. There must be the same number of branches from each of the nodes in an information set, and the branches are linked. An information set represents a single choice point at which the player does not have the option of making her choice depend on which of the nodes in the set she is choosing from.

Von Neumann and Morgenstern also defined a more abstract representation of their games – the *normal* or *strategic* form. This kind of representation abstracted away from the dynamic sequence of moves, specifying only the sets of alternative complete strategies that are available to each of the players, where a strategy is a conditional decision about which move to make at each of the choice points that might arise in the course of a play of the game. Any *strategy profile* (a sequence of strategies, one for each player) will determine a complete path through the tree, and so will determine the outcome of a play of the game when those strategies are played. So the strategic form of the game will include a function taking strategy profiles to the utility profile that is given by the terminal node of the path through the tree determined by the strategy profile.[2]

---

[2] Here is a formal definition: A game Γ in *strategic form* is a structure $\langle N, \langle C_i, u_i \rangle_{i \in N} \rangle$ where N is the set of players, C_i is the set of strategies available to player i, and u_i is player i's utility function,

Any game in extensive form will determine a unique strategic form representation, though different extensive form games might have the same strategic form. But it was argued that the strategic form includes all of the information about the game that might be relevant to the players' strategic reasoning about what moves it would be rational to make in playing the game, at least under the highly idealized assumptions that the theory makes about the cognitive capacities and rationality of the players. The thought experiment that motivated this conclusion goes like this: suppose, instead of playing out the game in real time, the players were to make *conditional* decisions, before playing the game, about what to do at each choice point that might arise, given the knowledge about how the game has proceeded up to that point that one would then have. Each player writes down his or her full strategy, and gives it to a referee, who then makes the moves determined, and announces the outcomes, and the payoffs determined. However rational players determine their choices (it was argued), the conditional decisions about what to do if a certain situation were to arise should be the same as the decision actually reached when the situation does arise.

We needn't suppose that the game will be played this way; strategies can be regarded as dispositions to choose, should the choice point be reached, or perhaps as implicit conditional intentions.

Just as Bayesian decision theory, in defining a decision problem, specifies the utility values that motivate a rational agent who faces that problem, so the definition of a game (in either extensive or strategic form) specifies the utility values that motivate the players. But in contrast with the way Bayesian theory models decision problems, the definition of a game does not include any specification of the beliefs or degrees of belief of the players about the moves that other players might make. There are some assumptions about what the players know and don't know about each other, but they are qualitative and informal. Players are assumed (by both the theorist and the players themselves) to be *intelligent*, and *rational*. What this means is that they know the structure of the game, including the utility values for the other players, and they know that all players will act with the aim of maximizing their own utility. The assumption that all players are rational is supposed to give each player a basis for predicting what other players will do, or would do if the game developed in a direction that it may or may not go. So while Alice has

---

taking strategy profiles (members of $C = \times_{i \in N} C_i$) to real numbers. The sets, N and each C_i, are finite. There are some complications in determining the strategic form of a game that contains chance moves, but we can ignore them here.

no control over what Bert will do if he gets the chance to act, and no direct information about how likely it is that he will choose one way or another, she does assume that he will act to maximize his own utility, given what he believes about her, and what he believes about her is that she will act to maximize her utility given what he thinks she believes about him. Probabilities enter the picture (on the traditional way of thinking about game theory) only in two ways: first, if the game includes nature as one of the players, the branches of the tree that represent her alternative moves will be assigned probability values that are given by the definition of the game. Second, the regular players have the option of using a randomizing device to choose their moves. That is, a player might play a *mixed strategy*, assigning probability values by choice to the different options available to him. Both kinds of probability were thought of as *objective* probability, and on Von Neumann's way of thinking about utility, the weighted average that is the expected utility is weighted by *objective* probabilities.

So the utility values given by the definition of a game are playing two roles in the players' reasoning about what actions and strategies to choose. A player's own utilities provide the motivation for her choices, while the utilities of other players provide a basis for predicting what choices they will make. To determine what Alice should believe about Bert, she puts herself in his shoes, considering what he would do, given his values, and given that he is doing what she is doing: she is simulating him, assuming that he is simulating her to predict what she would do. The threat of circularity is an obvious and familiar feature of game-theoretic reasoning, and it is not clear in the general case how such reasoning gets started, but the following, at least, is clear: each player knows that any information given by the definition of the game on which a rational decision might be based is available to all the players, so each knows that any reason that she has to act a certain way (at least any reason provided by the information given by the definition of the game) will be available for other players to use to predict her action. This seems to imply that *if* the game determines uniquely rational choices for the players, then each of those choices must be a best response to the others. This thought is the intuitive motivation for the notion of a *Nash equilibrium*, which is defined as a strategy profile for which each strategy in the profile is a best response to the other strategies in the profile.[3]

---

[3] Nash 1951. Nash equilibrium was a generalization of the kind of equilibrium that von Neumann and Morgenstern had defined.

The paradigm games with which Von Neumann and Morgenstern began were special cases in three ways: first, they were two-person games; second, they were *zero-sum* games, which means that they were games in which every benefit for one player was a cost for the other; third, they were perfect information games in the sense defined above. (Chess and tic-tac-toe are two familiar games satisfying these conditions.) If one focuses on a game of this kind, then the hypothesis that the structure of the game determines a uniquely rational solution[4] is plausible.[5] But as game theory developed, there was increasing emphasis on games that involve a mix of conflict and common interest (or even pure common interest), in games with many players, and in games of imperfect information. Any realistic situations of interactive decision-making (and idealized models of such realistic situations) that are of interest to social science applications are all of these more general kinds.

In the general case, some games will have no pure strategy Nash equilibrium, and with other games of interest there will be multiple equilibria that are not interchangeable. In games of this latter kind, rational players may have no motive for choosing an equilibrium strategy, even on the assumption that everyone knows that everyone is rational. Rational players must base their decisions partly on their beliefs and partial beliefs about what other rational players will do, and such beliefs cannot be determined by the players' knowledge of the structure of the game and of the utilities that provide the motivation for rational play, together with the assumption that everyone will choose rationally.

While in some games, there will be no pure strategy profile meeting the condition that every strategy in the profile is a best response to the others, if players have the option of choosing a mixed strategy (say by flipping a fair coin to determine which of two available strategies to play) then there will always be at least one Nash equilibrium – that is the fundamental theorem of game theory. But while this result has mathematical interest, and may be relevant to some applications of the mathematical theory, it is not clear that it, or the general notion of Nash equilibrium, is relevant to the question of what rational players in the decision situations modeled by game theory should do.

---

[4] Not strictly unique, since different strategy choices may in some cases determine the same outcome. But in games of this kind, if there are multiple Nash equilibrium strategy pairs, the strategies in them will all be *interchangeable*, meaning that every pair of Nash equilibrium strategies will itself be a Nash equilibrium strategy pair.

[5] Plausible given the extreme idealization, which ignores any computational limitations that real rational agents will have. Given the idealization, chess and tic-tac-toe are not very different from a strategic point of view – it's just that one is more complicated than the other.

It is not clear what motive a player could have for choosing to turn her decision over to a randomizing device, given the idealizing assumptions that the theory makes. And even when there are (multiple) pure strategy equilibrium profiles, it is not clear why the assumption of mutual rationality, or even of common knowledge of rationality, should motivate players to choose an equilibrium strategy.

2 Epistemic Game Theory

Classical game theory framed the main problem as the problem of defining the *solutions* to a game, which meant to identify the strategies that would be chosen by players who were presumed to know the structure of the game, and to know that all players would act rationally. To be rational meant to aim to maximize the player's utility, but the formal theory gave no precise definition of game-theoretic rationality. The epistemic conditions that were presupposed were just part of the informal commentary motivating the various solution concepts such as the Nash equilibrium and the refinements of it. The proposed solutions were to be assessed by their intuitive plausibility. But as interest in the wider class of games grew, theorists began to look for a more systematic and explicit formal account of the epistemic conditions that were presupposed, and for a more precise connection between the notions of rationality developed in Bayesian individual decision theory and the notion of rationality that was relevant to the evaluation of proposed solutions to games. The idea was to see a game as a *partial* specification of a sequence of decision problems, a specification that needed to be supplemented by a representation of the beliefs and partial beliefs of the players, including of the players' beliefs and partial beliefs about the beliefs and partial beliefs about each other. One supplements a game with a *model* of a particular playing of the game, drawing on the resources of the model theory for logics of knowledge and belief.[6] A model will specify the moves made by players at each choice point, and the moves they would have made at choice points that might have been reached, and it will specify what beliefs and credences the players have about each other at each point. Here is one way to specify such a model, for a game given in strategic form:

A model M for a game Γ is a structure $\langle W, \mathbf{a}, \langle R_i, P_i, S_i \rangle_{i \in N} \rangle$, where W is a state space, or a set of possible worlds, including all of the possibilities that

---

[6] See Battigalli and Bonanno 1999 for a survey of this way of interpreting game theory.

are compatible with the beliefs of any of the players; a is a member of W, the actual world of this particular playing of the game: the playing of the game being modeled. R_i is a binary doxastic accessibility relation for player i: xR_iy says that possible world y is compatible with what player i believes in possible world x; P_i is a normalized additive measure function that determines player i's partial beliefs in each possible world in the following way: the degree to which i believes proposition ϕ in world x, $P_{i,x}(\phi)$, is defined as the ratio of $P_i(\phi \cap \{y : xR_iy\})$ to $P_i(\{y : xR_iy\})$; S_i is a function taking possible worlds to strategy choices for player i: $S_i(x)$ is the strategy that player i plays in possible world x. *Propositions* (or to use the statistician's terminology, *events*) are subsets of W.

The specification of a player's credences and strategy choice is called the player's *type*, and the model is called a *type space*. So, using this jargon, a model specifies each player's type in each possible world. Since the game determines the utilities, and the player's type determines both what she does and what her credences are, the player's type will determine whether or not she is rational (in a given possible world) in the Bayesian sense – whether her strategy choice maximizes her expected utility.

The resources of a game model allow us to define a rich range of propositions: that player i is rational, that player i believes (with probability 1) that player j is rational, that it is common belief among the players that they all are rational, etc. One can then define the *class* of models in which a given proposition is true in the actual world of the model. For example, we can define the class of models in which it is common belief that all players are rational, and we can then determine what class of strategies are played in models meeting this (or any other) condition.

A *solution concept* is a class of strategies or of strategy profiles that (according to the concept) are recommended or required (under certain specified conditions) for rational players of the game in question. Using just the resources of classical game theory, one might give an algorithmic definition of a proposed solution concept, a set of strategy profiles that is claimed to include all and only those strategies that are intuitively reasonable for rational and intelligent players. The model theory provides a systematic way of evaluating any such proposal, making it possible to establish mathematically whether the strategies identified meet some specified epistemic condition. A model-theoretic definition of a solution concept is given by defining a class of models for the game – models that satisfy the specified conditions. The strategies that count as solutions are those that are played in some model meeting the specified conditions. For example, a strategy is *rationalizable* if it is played in a model in which there is common belief that all players are rational (where rationality is defined as maximizing expected utility).

Model-theoretic definitions of a solution concept correspond to algorithmic definitions in the way that model-theoretic definitions of satisfiability for a logic correspond to proof-theoretic definitions of consistency. There are characterization results for solution concepts analogous to soundness and completeness results for a logic. Rationalizability was given an algorithmic definition (a strategy is rationalizable if and only if it survives a process of iterated elimination of strictly dominated strategies), and it was proved that strategies meeting this condition are exactly those that meet the model-theoretic condition.[7]

So epistemic game theory is an application of Bayesian decision theory, but what version of Bayesian decision theory, causal or evidential, does it use? Epistemic game theorists have not addressed this issue explicitly, but games do represent causal structure, and do presuppose that causal independence is relevant to practical rationality in the way that causal decision theory does. The flow of information in an extensive form representation (where a player in some cases can make a choice depend on previous choices of other players) is reflected in the strategies defined for the more abstract strategic form of the game, and it is a crucial assumption of that kind of representation of a game that players' strategy choices are made independently: players do not have the capacity to make their choices depend on the choices of any of the other players. Dominance reasoning in such cases is taken for granted.[8]

Models for games in strategic form fit naturally into David Lewis's formulation of Causal Decision Theory (CDT). Lewis's theory represents a decision problem with a two-dimensional matrix, where the rows represent the alternative available actions, and the columns represent the alternative *dependency*

---

[7] This analogy is developed, and some theorems of this kind proved, in Stalnaker 1997.

[8] Newcomb-like problems are not considered by game-theorists, and it is generally assumed implicitly that the strategy choices of different players are both causally and epistemically independent, so one might argue that the notion of independence used in game-theoretic strategic form representations could be epistemic rather than causal. But in games with three or more players, it is important to note that the strategy choices of different players, while causally independent, need not be epistemically independent. If Alice is playing with Bert and Clara, even though Bert's strategy choice must be causally independent of Clara's, it is not assumed that those strategies are epistemically independent for Alice: she might have reason to take information about Bert's strategy to be evidence that is relevant to hypotheses about Clara's strategy. (The point is just about the notion of independence that is presupposed in a strategic form representation. This kind of epistemic correlation does not give us examples the kind of correlation that distinguishes the contrasting advice of CDT and EDT.)

hypotheses (which correspond, on a causal interpretation of Leonard Savage's theory, to what Savage called *states*). The expected utility of an action is defined as a weighted average of the value of an action, weighted by the credence of the dependency hypothesis. Here is Lewis's intuitive explanation of a dependency hypothesis:

> Suppose someone knows all there is to know about how the things he cares about do and do not depend causally on his present actions. If something is beyond his control, so that it will obtain – or have a certain chance of obtaining – no matter what he does, then he knows that for certain. And if it is within his control, then he knows that for certain.
>
> Let us call the sort of proposition this agent knows – a maximally specific proposition about how the things he cares about do and do not depend causally on his present action – a *dependency hypothesis* (for that agent at that time).[9]

In a two-person strategic form game, each player's strategy set is a dependency hypothesis set for the other player. More generally, in a many person game, player *i*'s strategy set will be her available actions, and the profiles of strategies for the other players will be player *i*'s dependency hypotheses. The model for the game will provide (for each possible world) player *i*'s credences for the dependency hypotheses, and the expected value of each strategy choice in the game will be as determined on Lewis's account.

The assumption that epistemic game theory presupposes causal decision theory may be disputed, and in any case it is not by itself an argument for causal decision theory. But it will help to see how game theory, on this interpretation, or on alternative interpretations, deals with some of the problems that have been raised for causal decision theory. That is what we will consider in the next three sections.

3 The Newcomb Problem and the Prisoners' Dilemma

The Prisoners' dilemma (PD) is probably the most famous game in the game-theoretic literature. It is thought to illustrate, in a simple and idealized way, a pattern that is present in a wide variety of real situations where interacting agents fail to achieve an outcome preferred by all of them, even though all act rationally. The name for the game comes from the following story that was told to illustrate the structure: two prisoners who are charged with

---

[9] Lewis 1981a: 8.

committing a crime together are interrogated separately, and offered the following deal (each knowing that his partner is being offered the same deal): if both confess, both will get a reduced sentence – say five years; if one confesses and the other does not, the one who confesses will get immunity and go free, while the other will get the maximum, say ten years. If neither confesses, conviction will be possible only for a minor charge, but they will both get a minor sentence, say one year. The prisoners care only about minimizing their own jail time, so each reasons as follows: if my partner confesses, I do better by confessing as well, getting five years rather than ten. But if he does not confess, I can get immunity by confessing, and so avoid jail time altogether. So either way, I should confess (realizing that my partner will do the same). The more abstract moral of the story is that an outcome that both players prefer to the one they foresee is unachievable for rational players acting independently.

David Lewis argued that there is not just an analogy between the Newcomb Problem and the PD: the PD just *is* a Newcomb Problem, or more accurately, a pair of Newcomb Problems, one for each prisoner.[10] He argued this by a slight modification of the PD story, keeping the payoff structure the same. Each prisoner will be presented with two boxes, a transparent one containing a thousand dollars ($K), and an opaque one. Each prisoner gets the contents of his opaque box no matter what, but each can choose whether or not to take the $K. The contents of each player's opaque box is determined by the choice made by the *other* prisoner: $M is put in player 1's opaque box if and only if player 2 declines the $K, and similarly for player 2's opaque box. There is no predictor in this story, but each player in this game can assume that he is very similar to the other player, and so each can assume that the other is very likely to make the same choice as he makes. So (Lewis argues) the prisoners' reasoning in this version of the PD should be the same as the reasoning of a player facing the original Newcomb Problem.

There were two ways in which Lewis thought that this conclusion helped to support CDT: first, the judgment that rational players should confess in the Prisoners' dilemma is widely accepted, and less controversial than the judgment that rational agents facing a Newcomb Problem should take the two boxes. So if the Prisoners' dilemma is a Newcomb Problem, the judgment about the game should lend support to the causalist's response to the decision problem. Second, it is sometimes argued that the Newcomb Problem is a bizarre and unrealistic scenario, and that intuitions about it can't be expected

---

[10] Lewis 1979.

to be a reliable guide to judgments about general principles of rational choice. But situations with the structure of the PD are widespread, and this game-theoretic structure helps to explain some real and important social phenomena.

But is Lewis's argument for the identification of the PD and the NP persuasive? In particular, is it plausible to treat another agent in a symmetrical game as, in effect, a predictor of one's own choice? I see a significant disanalogy, and while I think it is right that the proponents of CDT should see parallels in the way the agents should reason in these two stories, I doubt that the attempt to see the PD as a pair of Newcomb Problems has much dialectical force. To help explain why I think the original Newcomb Problem is relevantly different from Lewis's PD version, let me report on my reaction to the Newcomb Problem when I first encountered it more than forty years ago.

I thought, "It is obvious that one should take both boxes, and there is no puzzle about how a predictor could get it right 90% of the time, since surely almost everyone will make this choice, so the predictor can safely predict that everyone will, and be about 90% accurate." That was my first reaction, but while I retain my robust two-boxer intuitions, I soon saw that this analysis was a misunderstanding of the problem. For one thing, I was of course factually mistaken: a significant number – perhaps close to 50% – opt for one box, and in a full statement of the problem it should be specified that the predictor's accuracy rate is as high for one-boxers as it is for the two-boxers. If my original understanding of the problem were right, then even evidentialists would agree with the two-box choice; there would be no "why ain'cha rich?" argument, since no one would get the million.

But my first reaction to the Newcomb Problem would be the *right* reaction to Lewis's PD version of it. According to the way I was thinking of the problem, it is reasonable to think of the predictor as accurate because it is reasonable to think that there is a particular choice that I and others will all make, and that the predictor can figure this out. Were I a player in Lewis's PD game, and were I to assume that the other player, like me, will chose rationally, it would be reasonable for me to think that he would make the best response to my choice. Even if I haven't yet decided on my choice, I can determine that his best response will be to take the $K. Assuming he does this, there will be no money in my opaque box (though this doesn't matter to my choice). In any case, my rational response is to take my $K, and this outcome supports the belief that we will choose alike. So if I think of the other player as, in effect, a predictor of my choice, I will believe he is a very accurate predictor. On the other hand, if I give up the assumption that he is rational, allowing a significant chance that he will leave his $K, then I no longer have reason to think of him as (in effect) an accurate predictor of my choice.

The differences between the original Newcomb Problem and Lewis's New-combized PD are enough to allow an evidentialist who opts for the one-box solution to the original Newcomb to consistently also accept the game-theoretic analysis of the PD. (This is of course not an argument for one-boxing in the original Newcomb Problem – just an attempt to defuse an argument against it.)

Arif Ahmed, in his discussion of the analogy between the PD and Newcomb's Problem, considers a modified version of Lewis's PD, suggesting that what he calls the *Extended Dilemma* might provide a reason in support of the evidentialist's decision theory. In his variation, Alice's payoffs are exactly as in the standard PD: she gets the $M if and only if Bob does not take his $T. But Bob's payoffs are different: he "forfeits all the money if he makes a *different* choice from Alice."[11] The other crucial difference in the *Extended Dilemma* is that we assume that Bob knows Alice's choice before making his.

From the point of view of the game-theoretic representation of this scenario, the assumption that Bob makes his choice after learning Alice's is the assumption that he can make his choice conditionally on Alice's choice. So in the strategic form representation of the game, he has four strategies to choose from, rather than the two that are available in the simple PD. The matrix for the *Extended Dilemma* looks like this:

Table 1: "O/B" for example, means "take one box if the other player takes one box, take both if the other player takes both."

	Take one box	Take both boxes
O/O	($M,$M)	(0,$M)
O/B	($M,$M)	($K,$K)
B/O	(0,$K)	(0,$M)
B/B	(0,$K)	($K,$K)

Alice no longer has a dominant strategy, but Bob's conditional strategy (Take if she takes, leave if she leaves) weakly dominates all the others, and so is the only rational choice for him. Alice can predict this, and so her best response is to leave the $K in order to get $M rather than only $K. So the game-theoretic analysis clearly supports what all will agree is the reasonable

---

[11] Ahmed 2014a: 114.

solution. But must we think of the game-theoretic analysis as an application of *causal* decision theory? Ahmed argued that nothing in this reasoning "relied on [Alice's] choice being *causally* relevant to Bob's. It *might* be that in *Extended Dilemma* Bob learns what Alice does by seeing it. And in this case his choice *is* causally dependent on hers. But for all that I said about *Extended Dilemma*, Bob *might* be a Laplacean predictor who calculated Alice's choice in advance."[12]

I find it hard to take Laplacean demons seriously (I doubt they are physically possible, even given determinism), but even allowing them, I think it is still plausible to think of the game-theoretic analysis as correct, and as an application of Lewis's causal decision theory, since it is reasonable, in the strategic form of the game, to think of each player's strategy choices as dependency hypotheses for the other, where dependency hypotheses are defined as Lewis defines them. It is clear, whatever the explanation for Bob's knowledge of Alice's choice, that the four strategies given in the matrix are available to him. Suppose we were to play the game as in the story told by von Neumann and Morgenstern to motivate the strategic form: before learning what Alice chooses, we ask Bob to give us his strategy choice – how he proposes to choose, conditionally on learning what Alice will do. This choice is made independently of Alice's, as her choice is made independently of his.[13]

More needs to be said about Laplacean demons, soft determinism, and how a game should be represented when one of the players has access to determining causes of another player's future actions, but that will have to wait for another occasion.

4 Paradoxical Decision Problems

CDT observes that a decision may give the agent information of two different kinds about the state of the world that will result from her action. On the one hand, she will have beliefs and degrees of belief about the causal effects that her action will or might have; on the other hand, her action may tell her something she didn't know about herself, or about others, which would be true even if she had decided differently. For example, I may think that a friend knows me and my psychology better than I do, and probably knows better

---

[12] Ahmed 2014a: 115.
[13] The players' *strategy* choices are causally independent of each other, but because Bob's available strategies include strategies that make his choice conditional on Alice's choice, the choice he makes need not be causally independent of Alice's.

than I do what decision I will make in a situation I am still deliberating about, say about whether to accept a job I have been offered. When I finally decide to decline the job offer, I may come to believe as a result that this is probably what my friend predicted. All can agree that a decision may give one information of both of these kinds, but CDT holds that expectations about these contrasting kinds of information should play different roles in one's evaluation of the rationality of the action, and this can raise problems for the decision-maker who aims to follow the advice of that version of decision theory. A paradoxical decision problem is one where for each of the available actions, the hypothesis that that action is chosen gives one evidence that the facts are such as to make that decision irrational. The proponent of CDT faces the problem of what to do in such a situation.

Paradoxical decision problems were recognized in Robert Nozick's first presentation of the Newcomb Problem, and discussed in the first development of CDT. Allan Gibbard and William Harper used the old fatalist story of Death in Damascus to illustrate the problem. Death is surprised to see a certain man in Damascus, since he knows that he has an appointment with him the next day in Aleppo. The man tries to escape Death by rushing away to Aleppo. The man knows that Death will be looking for him at a certain predetermined place. If the place is Damascus, he ought to escape to Aleppo, but if it is Aleppo, he ought to stay in Damascus. Escape is futile, since the decision to go, or to stay, is decisive evidence that he should stay, or should go.[14]

This fantastical story is not itself a count against causal decision theory, since neither intuition nor the theory gives advice about it. Giving up seems the only reasonable response. But it has been argued that there are asymmetric examples of paradoxical situations where one choice (the one supported by Evidential Decision Theory) seems intuitively reasonable, even though CDT fails to deliver this conclusion.

To make this point, Andy Egan gave the following example, labeled PSYCHOBUTTON:[15] you are able to push a certain button that will result in the killing of all the psychopaths in the world. You think this would be a good option, assuming that you are not yourself a psychopath, but a bad option if you are. Your prior beliefs are that there is a very small chance that you yourself are a psychopath, but you believe that only a psychopath would push the button. So conditional on the hypothesis that you choose to push the button, you believe that it is very likely that you are a psychopath. It is clear intuitively, Egan argues,

---

[14] Gibbard and Harper 1978.
[15] Egan 2007. Egan's examples are discussed in Ahmed 2014a: ch. 3.

that you should refrain from pushing the button, which is what (assuming appropriate utility values) Evidential Decision Theory (EDT) prescribes.

What does CDT say about such cases? It seems to suggest that deliberation about them will lead to belief instability. Starting from your prior credence (that you are almost certainly not a psychopath), you tentatively decide that you should push the button. But since you know that if you make that choice you probably are a psychopath, you infer from your tentative decision that you are probably a psychopath, and so this is the wrong choice to make. So you reconsider and tentatively decide not to push the button. But then, on the hypothesis that you won't push the button, you infer that you *should* push it. Your beliefs remain in flux until you finally have to act, at which point, whichever choice you make, your beliefs will stabilize on the judgment that it was the wrong choice.[16]

So what advice does the causalist give? One short and not entirely satisfactory answer is that decision theory tells you what to do, given certain credences and utility values; it doesn't give advice to the agent who can't decide what credences to have. The causalist giving this response may add that any decision theorists, whatever version of the theory they defend, may face a problem of belief instability. The causal theory allows for a particular mechanism for generating belief instability in certain cases – a mechanism that will not apply for the evidentialist – but belief instability might arise for other reasons. Suppose (to give an autobiographically realistic example) the 2016 US presidential election is imminent, and I am deliberating about what to believe about the outcome. The pundits are saying different things, and the polls and meta-polls are yielding conflicting results. I am inclined to think that Clinton will probably win, but then I worry that wishful thinking is distorting my judgment, so I lower my credence. But then I worry that protective pessimism is biasing my thinking – that the awfulness of a Trump victory is making it seem more likely than the evidence justifies. My credence may continue to fluctuate until the results come in. Suppose I had to make a decision with a value that depended on the outcome of the election (perhaps whether to accept a job in Canada). How should I decide?

Can game theory throw any light on this problem? That theory does direct one's attention to examples (much more realistic than PSYCHOBUTTON) that

---

[16] One has to take this sketch of a process of deliberation with a grain of salt, since the idealizing assumption of logical omniscience that is made by decision theory (both EDT and CDT) abstracts away from any process of deductive reasoning. Even before making a decision, the agent can see what is coming.

may have the problematic structure. Suppose you are playing a complex purely competitive game against an opponent whom you believe to be better at the game than you. No matter what strategy you manage to come up with, you think your opponent will probably anticipate it. So any tentative decision to adopt the strategy gives one reason to think some other strategy would be better.

What should one do in such a situation? Epistemic game theory, with its extreme idealizations, is not much help, since like Bayesian decision theory, it assumes that players will have stable credences about the strategies of the other player, and can make the best response to those credences. Traditional game theory might suggest that one should adopt a mixed strategy, but despite the fact that there exist mixed strategy Nash equilibria, the following argument shows that, under the ideal assumptions of game theory, one can never have a motive to play a mixed strategy: on the one hand, if you assume that your opponent is playing a certain equilibrium mixed strategy, then any of the pure strategies that receive nonzero probability in your Nash mixed strategy is as good a response as any other, so there would be no point in taking the trouble to randomize. But on the other hand, if you assume that your opponent is *not* playing the Nash mixed strategy, then you will have a response that is better than your Nash strategy. In any case, mixed strategies won't provide any help with puzzles like the PSY-CHOBUTTON case, since one can easily stipulate that the decision to play a mixed strategy will give one reason to believe that one is a psychopath ("Only a psychopath would even take a chance on killing all the psychopaths").

But there is an intuitive argument for playing a mixed strategy in a competitive game in circumstances that are less than ideal, and I think this argument points to a way of thinking about paradoxical decision problems. Players with computational limits (which of course includes all real rational agents) don't select full strategies, but begin with a general open-ended plan. One might model the problem of choosing such a plan by a coarse-grained partition of available strategies; the initial problem is to choose between alternative classes of available strategies. To begin with a maximally coarse-grained partition, one might deliberate (at the start of a competitive game) about whether to empha-size offense (trying to exploit the weaknesses of one's opponent) or defense (trying to avoid being exploited by one's opponent). A mixed strategy in a competitive game models a purely defensive strategy: in choosing it, you give up any attempt to gain an advantage on your opponent, but make it strictly impossible for your opponent to gain any advantage on you. A player who believed that his opponent was a superior player (and for whom a mixed Nash equilibrium strategy was computationally accessible) might reason as follows: "If we both play an offensive strategy, I'll probably lose. If my opponent plays a defensive strategy, then an offensive strategy can't help, so either way I do best by playing defensively." If the inferior player looks at specific strategies, his beliefs about his opponent's strategies may become unstable, but the more general plan (play defensively) might be judged to be rational, in the Bayesian sense. And the advice, act cautiously – avoid risk – may be reasonable, in general, as a backup plan when one is faced with belief instability.[17]

This is not a solution to the problem of acting under belief instability, but it is a suggestion about how one might start thinking about such a solution.

5 Choosing Strategies vs. Choosing Actions in Sequence

As I have noted, game theorists argued early in the development of the theory that the strategic form of a game contains all the information about the game that is relevant to the players' strategic reasoning, under the idealizing assumptions that the theory makes. Players are assumed to revise their beliefs by conditionalization, and to assume that others will do the same. And it is assumed that conditional decisions about what to do in a situation that might or might not arise will be the same as the decisions that would be made if and when the situation does arise.

This thesis was controversial, but in epistemic game theory, it can be made more precise, and it can be proved (oversimplifying a bit) that a strategy choice in a model of the strategic form of a game is rational if and only if each decision prescribed by the strategy would be rational in the situation that would arise in the corresponding model for an extensive form version of the game.[18] Under less ideal conditions than those made by game theory, the choices made by a rational player in a real-time playing out of a game may differ from the choices determined by the full strategy that the player would make, and there are some dynamic variations on puzzle cases discussed in the debates between CDT and EDT. Let's see if the game-theoretic representations can clarify what is going on in such cases.

---

[17] See McClennen 1992 and Gibbard 1992 for interesting discussions of mixed strategies and belief instability.

[18] See Stalnaker 1999 for the statement and proof of this theorem. The argument requires a slight strengthening of the notion of rationality as maximizing expected utility, but it is a modification that is independently motivated, and independent of issues on which Causal and Evidential decision theory disagree. Strategy choices must be *perfectly* rational, which requires that in case of ties in expected utility, rational players should also maximize conditional expected utility on the condition that some proposition with prior credence 1 is false.

There is a family of examples of sequences of two decisions discussed in Arif Ahmed's book, which he calls "insurance" cases.[19] Each can be seen as a one-player game with a certain strategic form, and the prescriptions of game theory (and CDT) seem highly counterintuitive. Here are two examples, each consisting of a sequence of two decisions, the second being an opportunity to bet on the outcome of the first:

Psycho-insurance: The first decision has the structure of PSYCHOBUT-TON, with a predictor, and the assumption that she is known to be 95% accurate. The second decision is an opportunity to bet that the predictor predicted correctly, or to decline the bet. Here are matrices giving the utilities for the different outcomes for each decision, and a matrix representing the strategic form of the two-step game.

Table 2: Stage one.

	Predict O1	Predict O2
O1	−2	+2
O2	0	0

Table 3: Stage two.

	Predictor correct	Predictor wrong
Bet	+1	−3
No bet	0	0

Table 4: Strategic form.

	Predict O1	Predict O2
O1,B	−1	−1
O1,~B	−2	+2
O2,B	−3	+1
O2,~B	0	0

How should one plan one's strategy? In the strategic form, the two strategies that include accepting the bet are each dominated by a strategy that includes rejecting the bet, but it seems obvious that accepting the bet is the rational

---

[19] The psycho-insurance case is discussed on pp. 226ff, and Newcomb-insurance on pp. 201ff. The two examples are used by Ahmed to make somewhat different points.

choice at that point, since one's belief that the predictor is 95% accurate remains stable throughout. What should one make of this situation?

CDT says, about the simple PSYCHOBUTTON, and so about step one of this problem, that whatever you do, you will have beliefs about dependency hypotheses that require the opposite choice. In a dynamic choice situation where you believe (at a later stage) that you made the wrong choice (at an earlier stage), you should make the best of it from then on, and that may result in an overall strategy that is strictly dominated. Even if you had to turn in your strategy in advance, you will recognize that the problem has this form, and so should accept the rationality of the second choice prescribed by a strictly dominated strategy.

The second insurance problem is also a case where the intuitively reasonable choice is strictly dominated, but the explanation is somewhat different.

Newcomb-insurance: The first decision has the structure of a standard Newcomb Problem. The second decision is, again, the opportunity to bet on the success of the predictor. This time, you can bet that she was right, or that she was wrong. Here are the matrices for the two decision problems, and for the strategic form of the combination:

Table 5: Stage one.

	Predict OB	Predict BB
OB	100	0
BB	110	10

Table 6: Stage two.

	Predictor correct	Predictor wrong
Bet correct	25	−75
Bet wrong	−25	75

Table 7: Strategic form.

	Predict OB	Predict BB
OB,Bc	125	−75
OB,Bw	75	75
BB,Bc	35	35
BB,Bw	185	−15

As in the original Newcomb Problem, it is assumed that the agent believes, with credence .9, that the predictor predicts the choice at stage one correctly.

Again, the two strategies that include betting that the predictor was correct are each strictly dominated, but again it seems obvious that the rational choice, at stage two, is to accept the bet.

This case is different in that in the simple Newcomb case, the causal decision theorist thinks there is a uniquely rational choice, but when the betting option is added, this gives the player a motive that he does not have in the simple Newcomb Problem to try to defeat the predictor. The result is that the overall game represented in the strategic form has a paradoxical character similar to that of the simple PSYCHOBUTTON case. If you in fact choose two boxes at stage one, you then will believe that the optimal *strategy* was to choose one box, and then bet against the predictor. On the other hand, if you choose one box at stage one, then you will believe that the optimal overall strategy would have been to choose both boxes, and then bet against the predictor. But at stage two, you no longer have the option of choosing the strategy that you then believe would have been optimal. At that point, you should make the best of it, and bet that the predictor was right.

As noted at the start of this section, under the idealizing assumptions made by game theory, *strategies* are rational if and only if they everywhere prescribe *choices* that are rational, but now we are saying that in the insurance cases, choices prescribed by a dominated strategy (which therefore cannot maximize expected utility) are rational. The idealizing assumption that fails in these cases is that credences evolve by conditionalization. In the agent's prior situation, before choosing an action or a strategy, credences are unstable, and whichever action is chosen at stage one, the credences that rationalize it will be judged, at stage two, to be the wrong ones to have had. The credences at that stage will therefore not be those that result from conditionalizing on those of the earlier stage.

There remains the problem for CDT about what to say about the paradoxical decision problems, but I don't think there is an additional problem about what to say about these dynamic sequences that involve paradoxical decisions.

What does EDT say about choice sequences and their strategic forms? Here is another dynamic variation on the Newcomb Problem: *Newcomb-reveal*:[20]

---

[20] *Newcomb-reveal*, or similar cases, have been discussed in many places, beginning with Gibbard and Harper 1978. See Wells, forthcoming, for a recent discussion of the problem, which Wells labels "Viewcomb." He constructs a more complex variation on this problem and argues that it provides a sound "why ain'cha rich?" argument against EDT.

the situation is like the standard Newcomb Problem, with a predictor with a 90% accuracy rate. The difference is that both boxes will be transparent, so that the agent will know, before making his choice, whether there is $M in what we will call the big box, or nothing. At least that is the default case. Before the boxes are presented, the agent has the option, for a fee of $10K, of covering the big box, so that the problem at stage two will be just like the original Newcomb Problem. The predictor makes her prediction before learning what the agent chooses at either stage one or stage two.

EDT agrees with CDT that if the agent knows whether the money is in the box before making his choice, he should pick both boxes, either way. So the evidentialist can predict that if he leaves the big box transparent, he will probably get just $K, whereas if he opts to cover it, and chooses one box, he will probably get the $M (minus the $10K fee) There is a 10% chance of getting nothing for the $10K fee, but it is worth the risk.

But suppose the game were played as in von Neumann and Morgenstern's thought experiment by committing to a strategy in advance. The strategy, "leave the big box uncovered, but reject the $K whether there is money in the big box or not" maximizes the evidentialist's expected utility. It seems that the strategy choice, or at least the thought experiment for implementing it, is a way of binding oneself to a choice you would not make if you were not bound. But the two-stage decision problem includes no binding opportunity. It seems that the agent who is disposed to leave the big box uncovered, but to reject the $K in any case, is not the rational evidentialist, but a resolutely irrational evidentialist. Such irrational agents get even richer than the rational evidentialist, when faced with a problem like *Newcomb-reveal*.

It is reasonable, I have argued, to think of the game-theoretic representation of decision problems as an application of Causal Decision Theory. The game framework does not provide decisive answers to the problems that the causalist faces, but I have tried to suggest that it does provide some tools that may help to clarify those problems.[21]

[21] Thanks to Arif Ahmed for very helpful comments on two previous drafts of this chapter, and to Ian Wells and Jack Spencer for discussion of the issues at stake.

10 Diagnosing Newcomb's Problem with Causal Graphs

Reuben Stern

1 Introduction

You stand before two boxes. One is transparent and contains $1,000. The other is opaque. You have a choice. You can *one-box* – i.e., take the contents of the opaque box – or *two-box* – i.e., take the contents of both boxes. The game is set up such that the contents of the opaque box always depend on the earlier prediction of a remarkably successful predictor. If the predictor predicts that you will one-box, she places $1,000,000 inside the opaque box. If she predicts that you will two-box, the opaque box is empty. Should you one-box or two-box?

This is *Newcomb's Problem* (NP). As the readers of this book know, the answer to this question is controversial. Evidential decision theorists typically say that you should one-box because one-boxing is strong evidence for $1,000,000 being in the opaque box. Causal decision theorists reply that you should two-box because your choice exerts no causal influence over whether the predictor puts $1,000,000 in the opaque box (since the present does not cause the past), thus ensuring that two-boxing makes you $1,000 richer than you would have been had you one-boxed.

Who is right? Since there are competing intuitions about NP (even when carefully scrutinized), it is imprudent to focus exclusively on NP as we answer this question. Rather, we should focus on the underlying rationales that fuel each response. Traditionally, each camp has adopted its own version of decision theory in order to deliver its favored recommendation. Evidential decision theorists argue that agents should choose whatever action is evidence for the best outcome (i.e., whatever action maximizes conditional expected utility),[1] while causal decision theorists often argue that agents should deploy the unconditional probabilities of non-backtracking counterfactuals as they

---

[1] The conditional expected utility of a given action is a weighted average of utility across possible outcomes, where the weights are the agent's credences in the outcomes given that she takes the action.

calculate expected utility.[2] As such, the dispute is usually thought to turn on which of two irreconcilable decision norms is more intuitive. If we go the former route, we allow non-causal backtracking probabilistic dependencies to affect expected utility, and thereby secure the evidential decision theorist's intuition that subjects should one-box. If we go the latter route, we disallow backtracking probabilistic dependencies from affecting expected utility, and thereby secure the causal decision theorist's conviction that subjects should two-box.

Though walking through the details of these competing decision theories can help us to see the implications of siding with one-boxing or two-boxing (since we can trace where their recommendations come apart in cases other than NP), it does little to move the dispute forward.[3] This is because typical statements of Evidential Decision Theory and Causal Decision Theory simply codify the intuitions that people immediately express upon confronting NP, and thereby fail to give either camp ammunition to convince members of the other to switch sides. The causal decision theorists have their intuitions, the evidential decision theorists have theirs, and the result is loggerheads.

Is there any way out of this deadlock? If there were some way of viewing the dispute as rooted in something other than which norm is more intuitive, then perhaps inroads could be made. For if either position could be shown to implicitly rely on some hidden assumption about something other than the (ir)relevance of non-causal correlations (or backtracking dependencies), then philosophers could potentially make progress on NP by shifting their focus to the status of the hidden assumption.

Luckily, there is a way of viewing NP as depending on such an assumption. When viewed through the lens of the most promising framework for under-standing the relationship between causal relevance and probabilistic rele-vance – i.e., the graphical approach to causal modeling developed by, e.g., Pearl (2009) and Spirtes, Glymour, and Scheines (2000) – the dispute between one-boxers and two-boxers appears to turn on how exactly agency should be causally represented, rather than on which of two irreconcilable decision

---

[2] See Lewis 1981a; Gibbard and Harper 1978.
[3] When philosophers argue that either theory yields untoward results when applied to cases other than NP, both camps tend to respond by explaining either (i) why the rationale that vindicates their preferred Newcomb response does not (when carefully specified) vindicate the counterintuitive verdicts, or (ii) why we should question what initially seems intuitive in these alleged counterexamples. See the literature on "medical Newcomb Problems," as well as the literature in response to Egan 2007 and Ahmed 2014b.

norms is more intuitive.[4] For example, in my (2017) paper, "Interventionist Decision Theory," I use causal graphs to develop a decision theory that secures causal-decision-theoretic verdicts only by requiring that agents are represented as *intervening* to make themselves act (or, in the case of NP, by representing agents as acting through means that are exogenous to the causal system that explains the correlation between the predictions and the actions).

In this chapter, I go beyond the standard treatment of these issues by considering various assumptions about choice that one might make within the causal modeling framework (including some that correspond to representing the agent as intervening, but also others) and highlight their implications for NP by considering variants of NP where their recommendations come apart. Though I do not defend any particular assumptions that I present, I consider how they relate to substantive views about choice that are defended in other contexts.

Since philosophers typically have opinions about the nature of choice that are formed independently of their opinions about NP, my hope is that this exercise can help to move the NP dispute forward. For example, after reading this chapter, two-boxers may find that graphical causal models vindicate their position only given a conception of agency that they are themselves unwilling to accept (and they may correspondingly reconsider their opinion of two-boxing). Or one-boxers might do the same. My hope here is thus not to *solve* NP (by showing that one-boxing or two-boxing is the solution), but rather to *diagnose* NP by showing how its solution can be seen to depend on the answer to the difficult question of how genuine choice should be represented.

In order to accomplish this, I utilize a norm of rational choice that is neutral with respect to how agents must represent themselves when making choices. This allows for the assessment of a specific view about the nature of choice by strengthening the neutral rule so that it does not apply to any scenarios ruled out by the specific view. My strategy for arriving at the neutral rule is to generalize my (2017) interventionist decision theory so that it requires agents to calculate the expected utility of x-ing in correspondence with *intending* to make oneself x (rather than in correspondence with

---

[4] To my knowledge, Meek and Glymour (1994) were the first to notice this. Though I disagree with some of the details of their analysis, I am in agreement with their key insight that the nature of the disagreement between evidential decision theorists and causal decision theorists is *not* best treated as a disagreement about the fundamental normative principles that govern rational choice, but rather as a disagreement about the nature of choice. Indeed, in this chapter, my aim is to determine how different views about the nature of choice yield different verdicts for NP when plugged into one and the same norm of rational choice.

intervening to make oneself *x*), and leaving it open whether intentions must be interventions, or must satisfy some other constraints.

2 Intervening in Newcomb's Problem

Given that we want to use causal graphs to think about NP, our first task is to draw a causal graph of NP. This means searching for a directed acyclic graph (DAG) that captures the causal features of NP that the subject has reason to believe. The outcome in NP – i.e., the amount of money received by the subject – causally depends on both the predictor's prediction and the subject's action. These are correlated. (If they were not correlated, then the predictor could not be successful.) How do these facts help us identify a plausible DAG?

The Causal Markov Condition (CMC) – i.e., the axiom of the graphical approach to causal modeling that captures the sense in which causes "screen off" their effects – entails something very close to the following variant of Reichenbach's principle of the common cause (PCC).

> PCC: If variables F and G are correlated, then either F (directly or indirectly) causes G, G (directly or indirectly) causes F, or F and G are (direct or indirect) effects of a common cause.

It does not seem that the predictor's prediction causally depends on the subject's action because the predictor's prediction temporally precedes the subject's action. It also does not seem that the subject's action causally depends on the predictor's prediction; NP depicts the predictor's prediction as a remarkably successful indicator of the subject's action, but not as in control of the subject's action. So the PCC tells us that we should search for some common cause. It is plausible that there is some fact that both informs the predictor's prediction and causes the subject's action. For example, the subject's brain states prior to acting may influence both the predictor's prediction and the subject's action. Allow us to represent the subject's earlier brain states with B, the subject's action with A, the predictor's prediction with P, and the monetary outcome with O.[5] The following DAG seems to give us some handle on NP.

---

[5] There may be other suitable candidates for the common cause of P and A, but nothing I say in what follows depends on the particular common cause that we choose.

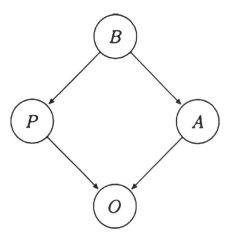

Figure 1

In treating Figure 1 as a DAG, I assume that each of its variables is a function of its parents (its most immediate causal predecessors) and its error term (if it has one).[6] Though it is often assumed that each variable has its own error term, it is acceptable to omit the representation of error terms from graphs. This is because error terms are (by stipulation) causes of only single variables, not common causes. So their omission does not induce any spurious correlations.[7] Below is a DAG in which the presence of error terms for each variable is made explicit.

What does Figure 2 tell us about NP? The CMC not only entails the PCC, but also that some variables must be probabilistically independent given a DAG. Pearl (2009: 16–17) neatly summarizes these implications in the graphical terms of **colliders** and **d-separation**.

> **collider**: A variable is a *collider* along a path if and only if it is the direct effect of two variables along the path. (So G is a collider along $F \rightarrow G \leftarrow H$ but not along $F \leftarrow G \rightarrow H$ or $F \rightarrow G \rightarrow H$.)[8]
>
> **d-separation**: A path between two variables, X and Y, is *d-separated* (or blocked) by a (possibly empty) set of variables, Z, if and only if
> i. the path between X and Y contains a non-collider that is in Z, or

---

[6] A variable is a partition whose values form a set of mutually exclusive and collectively exhaustive possibilities. For example, propositions can be represented as variables that take one value when true and another when false.

[7] For the purposes of this chapter, a cause C can be justifiably omitted from the DAG if and only if C is not a common cause of any variables that appear in the DAG.

[8] A path is just a consecutive string of arrows, no matter their direction. So in Figure 1, the paths between B and A are $B \rightarrow A$ and $B \rightarrow P \rightarrow O \leftarrow A$.

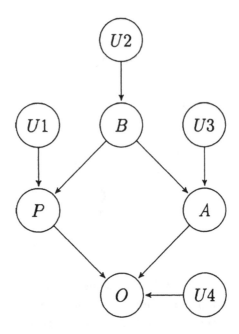

Figure 2

ii. the path contains a *collider*, and neither the collider nor any descend-
ant of the collider is in Z.

Given the CMC, if every (undirected) path between a pair of variables in **V**
is *d-separated* by Z (according to a given DAG), then the pair of variables
must be probabilistically independent of each other conditional on any
assignment of values over Z.[9] This means, for example, that every omitted
cause represented by the error term, U_3, must be probabilistically independent
of every endogenous variable that is neither A nor causally downstream of A
because every path between U_3 and every such variable is blocked when Z
denotes the empty set.

With d-separation in hand, we can understand the effect of *intervening* on
a causal system to set a variable to a particular value. Following Spirtes,
Glymour, and Scheines (2000), allow the *intervention* on A to represent some
justifiably omitted cause of A that can be exploited to set A to a. More
specifically, allow the intervention variable, first, to represent some cause of
A that is neither included in **V** nor a direct cause of any variable in **V** other
than A, and, second, to be partitioned such that it contains (i) a value on

---

[9] **V** denotes the set of variables over which a DAG is defined.

which one can condition to deterministically set A to a for every a in A and (ii) a value that corresponds to *not* intervening on A (i.e., to allowing the probability distribution over A to be determined by A's causes in $\mathbf{V}$).

Since the intervention on A must be a justifiably omitted cause of A, the intervention must be probabilistically independent of A's non-descendants in $\mathbf{V}$.[10] This means that the causal character of interventions entails that intervening to make oneself one-box (or two-box) is not correlated with the predictor's prediction, and that one can therefore use evidential norms to deliver causal-decision-theoretic verdicts simply by modeling the agent as intervening on the causal system depicted at hand. That is, if we model the NP subject as intervening, then she should two-box because the predictor's remarkable success has no bearing on whether she *intervenes* to make herself one-box or two-box (even though the predictor is remarkably successful at determining whether subjects generally one-box or two-box).[11] As Pearl (2009: 109) puts it, interventions "*change* the probabilities that acts normally obey."

What decision theory delivers these results? In another (2017) paper, I argue that interventionist decision theory (IDT) does a better job than its competitors at using causal graphs to deliver causal-decision-theoretic verdicts.[12]

INTERVENTIONIST DECISION THEORY (IDT)

Choose the act that maximizes expected utility when calculated as follows:

$$U(x) =^{df} \sum_K P(K) V(do(X = x), K).$$

Here, $do(X = x)$ represents setting the action variable to x by intervention, and the elements of the K partition are causal hypotheses over the variable set at hand, where causal hypotheses are construed as ordered pairs of DAGs and objective chance distributions (or sets of objective chance distributions).[13]

---

[10] Because A is a collider on every path from the intervention variable to A's causal predecessors, the intervention variable must be probabilistically independent from A's causal predecessors. Because the intervention variable is a justifiably omitted cause, it must be probabilistically independent of every other variable that is not causally downstream from A.

[11] NP is sometimes described such that the predictor's success is defined over the token subject, and not the type of subject who confronts NP. As I see things, this amounts to stipulating away the possibility of intervention.

[12] In this other 2017 paper, I argue that the causal decision theorist should prefer this definition of expected utility to any allegedly partition-invariant definition in terms of $P(y|do(x))$.

[13] In the 2017 paper, I maintained that causal hypotheses are ordered pairs of DAGs and *sets* of objective chance distributions in order to allow for non-singleton sets of chance distributions to count as causal hypotheses. I did this primarily because I did not see any reason to rule out the

Though IDT may sound complex, the basic idea is simple. According to IDT, the agent should determine the expected utility of x-ing by, first, spreading her subjective probability distribution across the possible ways that the world could causally be (i.e., across the K partition), and, second, using those probabilities to calculate a weighted average of the utilities she attaches to intervening to make herself x when each causal hypothesis is realized. My reason for construing causal hypotheses in terms of ordered pairs of DAGs and objective chance distributions is simply that being n confident in a causal hypothesis over some variable set corresponds to being n confident in the conjunction of a DAG and objective chance distribution that meets the constraints implied by the DAG, where the distribution specifies the underlying nature and strength of the dependencies qualitatively represented in the DAG.[14] The agent can thus determine the expected utility of x-ing by calculating a weighted average of the utilities that she attaches to the worlds that result from intervening to make herself x when each of the live causal hypotheses (or ordered pairs) is realized, where the weights correspond to the subjective probabilities that she assigns to each of the live causal hypotheses.

In NP, since the chance of intervening to make oneself one-box or two-box must be independent of the predictor's prediction in every chance distribution that is consistent with Figure 1, the agent knows that two-boxing will beat one-boxing no matter which causal hypothesis is realized, and her subjective probabilities over the causal hypotheses therefore play no role in determining what she should do – i.e., two-boxing *dominates* one-boxing. Of course, this is true only because IDT requires that we model the agent as intervening to make herself act, and it is not yet clear whether or why this might be reasonable.

possibility of this kind of indeterminacy within a causal hypothesis, but also because I thought that doing so might help render IDT compatible with the Spohn (1977) and Levi (1997b) thesis that "deliberation crowds out prediction." I no longer think this renders IDT compatible with the Spohn/Levi thesis, but remain open to this kind of indeterminacy. For ease of exposition, I speak almost entirely in terms of (single) objective probability distributions in what follows.

14   According to this construal of chance, chances reflect causal tendencies that do not change over time. This means that if I told you that a ball was randomly selected from a fair 100-ball urn containing either 40 red balls or 60 red balls, and then asked what chance distributions you were entertaining, it would be reasonable to mention distributions according to which the probability of its being red is .4 or .6, as well as distributions according to which the probability of its being red is between .4 and .6 (e.g., if you entertain causal hypotheses according to which there is reason to expect there to be 40 balls with some non-extreme probability). But even though the ball has already been drawn, it would not be reasonable to include distributions according to which the probability of its being red is 1 or 0. This places the operative notion of chance at odds with those according to which chances of past events must be 0 or 1.

3 Generalizing Interventionist Decision Theory

Given the axioms of the graphical approach to causal modeling, we can develop a decision theory that uses evidential reasoning to get causal-decision-theoretic verdicts, but only by stipulating that agents are modeled as intervening. Were we, for example, to deploy the machinery of IDT, but apply the value function to worlds in which $X = x$ (rather than $do(X = x)$), the results would be those associated with Evidential Decision Theory. That is, though intervening to make oneself one-box (or two-box) must be uncorrelated with the predictor's prediction in every candidate chance distribution, one-boxing (or two-boxing) itself must be strongly correlated with the predictor's prediction in every live chance distribution (since we know that the predictor is remarkably successful). So if the agent were to first spread her subjective probability distribution across the candidate chance distributions (the possible rates of success that she thinks the predictor may have), and then use those probabilities to calculate a weighted average of the utilities she attaches to x-ing (rather than intervening to make herself x) when each distribution is realized, she would discover that she should one-box.[15]

Should NP subjects be represented as intervening? It cannot be true that every NP subject actually intervenes (since this is inconsistent with regarding the predictor as remarkably successful). So it seems that we need to assess whether there is something special about deliberation or intentional action that justifies representing agents as intervening even when they are not. More generally, we need to consider how different views about the nature of agency interact with a decision-theoretic framework that utilizes causal graphs. In order to accomplish this, we must generalize IDT so that it applies to contexts where agents are not represented as intervening.

We can revise IDT to accommodate contexts where an agent is less than certain that she is intervening by applying the value function to the *intention* to x and leaving it open whether the intention is an intervention – i.e., by analyzing the expected utility of x-ing as follows, where i_x is the value of the intention variable that corresponds to intending to x (or settling on x-ing).[16]

---

[15] Again, by stipulating that the chance distributions encode the possible rates of predictive success, I must depart from any view according to which the chances of past events must be 0 or 1. If the chance that the predictor predicted one-boxing (or two-boxing) was 1 (or 0), then, given standard treatments of correlation, P could not be correlated with A.

[16] Some may wonder why GIDT asks agents to opt for whatever *act* maximizes expected utility rather than whatever *intention* maximizes expected utility. I am simply following the lead of IDT (which itself follows the lead of Pearl 2009 and Meek and Glymour 1994) since it is in terms of $U(x)$ rather than $U(do(x))$.

GENERALIZED INTERVENTIONIST DECISION THEORY (GIDT)

Choose the act that maximizes expected utility when calculated as follows:

$$U(x) \overset{df}{=} \sum_K P(K)V((I = i_x), K).$$

GIDT generalizes IDT so that it applies to contexts where the agent is less than certain that she is intervening because "the intention to x" is construed broadly enough to refer to both the intervention to make oneself x and the endogenous intention to x – e.g., such that it refers to I in each of the below DAGs of NP in Figure 3 and Figure 4. Since not every path between P and I is d-separated in Figure 4, GIDT is weak enough to range over scenarios where the agent's intention *is* evidentially relevant to the predictor's prediction. Of course, if agents should be represented as intervening in every decision-making context, then GIDT effectively reduces to IDT.[17] But unlike IDT, GIDT does not stick its neck out with respect to how the agent's intention should be represented, and thereby provides a setting in which the implications of assuming different constraints on how intentions should be represented can be explored.[18]

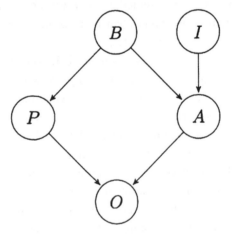

Figure 3

---

[17] This is because all of the doxastically possible worlds that correspond to *intending* to one-box (or two-box) are then worlds in which she is intervening, even if only some small portion of the doxastically possible worlds in which the agent one-boxes (or two-boxes) *simpliciter* are worlds in which she is intervening.

[18] One might reasonably worry about applying the value function to worlds in which the agent *intends* to x rather than worlds in which the agent x's because it seems to depict the agent as choosing what to choose (or intend), rather than as choosing what to do. This shift in focus is

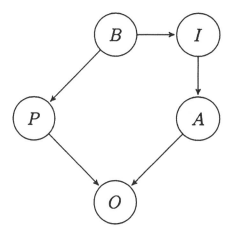

Figure 4

4 Intentions as Interventions

With GIDT in hand, we stand in a position to consider the implications of assuming different views about the nature of choice through the lens of causal graphs. Consider the view according to which every genuine intention must be represented as an intervention (i.e., IDT).

Why think that agents must be represented as intervening? I suspect that the primary reason that people are attracted to this view is that it captures the intuition that the outcome of an agent's choice must be *up to the agent* in order for the choice to issue a genuine intention. There are two senses in which regarding intentions as interventions accomplishes this. First, if intentions must be interventions, then genuine intentions are causally autonomous in the sense that they are not caused by anything in the model under consideration. (Relative to the model at hand, the agent appears to be the ultimate source of what she does.) Second, if intentions must be interventions, then the agent is granted what Velleman (1989) has called "epistemic freedom" since her beliefs about how she will choose are not constrained by other evidence that she might have about what she is likely to do.[19] This is because the causal character of an intervention entails that the intervention to act

---

shared by every decision theory where the value function is applied to *do(x)* – including Hitchcock (2016), Meek and Glymour (1994), and Pearl (2009) – because interventions are *not* the acts themselves, but rather causes of the acts.

[19] Joyce (2007) channels Velleman (1989) in his defense of the *evidential autonomy thesis* – i.e., the thesis that "a deliberating agent who regards herself as free need not proportion her beliefs about her own acts to the antecedent evidence that she has for thinking that she will perform them."

must be uncorrelated with any of the variables in the DAG that are not causally downstream from the action variable (which may correspond to the variables that the agent could have evidence about, given that they are the variables whose values are determined prior to the formation of the agent's intention). Thus, when considered through the lens of causal graphs, the epistemic freedom of the agent is secured by her causal autonomy.

Though some have intuitions that we must be causally autonomous in order to make genuine choices, there is reason to worry that this requirement leads to an unacceptable skepticism about choice. Indeed, when philosophers argue that an agent freely wills her action only if she is the ultimate source or ultimately responsible for her action, this typically results in thinking that we freely will our actions only if the universe is indeterministic, or – more drastically – that we do not freely will our actions no matter the status of determinism.[20] Though the details of these conceptions of free will evade us here, it is easy to see that they lead down the path to skepticism. When I settle on having a peanut butter sandwich for lunch, it is clear that I do so partially because I've been exposed to peanut butter, because there is peanut butter on the shelf, because George Washington Carver invented peanut butter, etc.[21] So it is hard to see how my intention to reach for the peanut butter could qualify as the *ultimate* source of my doing so (or as causally autonomous), even if it is *a* source of my doing so.[22]

Perhaps it is reasonable to be skeptical about free will,[23] or to be no more confident that we have free will than that the world is indeterministic, but *surely* we often make choices. That is, regardless of whether I *freely* choose to

---

[20] Kane (1996) argues that free will is incompatible with determinism and that we must be ultimately responsible for our x-ing in order to freely will it. Strawson (1986) argues that we lack free will (no matter the status of determinism) from the premise that we must be the ultimate source of our actions in order to freely will them.

[21] According to some analyses of what it is to be causally autonomous in the relevant sense, the agent's intention need not be uncaused. Rather, the agent's intention must not be "bypassed" in the sense that it screens off every other cause of what the agent does. As these examples make clear, even this weaker sense of causal autonomy is threatened in normal cases since, e.g., my intention to eat peanut butter does not screen off the relevance of whether there is peanut butter on the shelf to whether I eat peanut butter (since whether I eat peanut butter still depends on whether there is peanut butter for me to eat even given that I intend to eat peanut butter).

[22] Astute readers will notice that the intervention to act must be causally autonomous only relative to the variables under consideration. But if IDT is to be applied in any genuine decision-making context, then its defender is committed to the strong thesis that agents should be modeled as causally autonomous relative to every admissible variable set, which seems to reduce to the view that intentions must be causally autonomous *simpliciter*.

[23] See Strawson (1986) for arguments as to why this may not be as troubling or counterintuitive as it first seems.

peanut butter, I clearly make a choice – i.e., I *make up my mind* insofar as I don't know whether I will reach for the peanut butter as I deliberate about whether to do so, but know that I will reach for the peanut butter if I settle on doing so. So it seems wrong to think that intentions must be modeled as interventions for the express reason that doing so is consistent with modeling intentions as causally autonomous.

Does this spell doom for IDT? No. Even if intentions need not be causally autonomous in order to qualify as genuine intentions, it may be that when evaluating the rationality of an intention, we should evaluate it as though it *were* causally autonomous. This is because it may be that agents must momentarily represent themselves as causally autonomous as they make choices (even if they are not). Many philosophers have defended this thesis on the grounds that agents cannot represent their choices as *open* in the relevant sense unless they represent themselves as causally autonomous,[24] and, if this is right, then it is plausible that their *rationality* should be determined given their representation of themselves (since their rationality should be determined *given* their beliefs – including those about their causal relation to things).

No matter whether intentions must be represented as causally autonomous,[25] it is clear that doing so yields causal-decision-theoretic verdicts when plugged into GIDT for the simple reason that GIDT reduces to IDT when every intention is represented as an intervention.[26]

5 Deciding without Causal Autonomy or Epistemic Freedom

Though some may argue that an agent must represent herself as causally autonomous, this is not the only way to account for the sense in which the

---

[24] Ramsey (1978), Joyce (2007), Price (1992), and Spohn (2012) are prominent examples in the decision-theoretic literature.

[25] Spohn (2012) thinks that intentions must be represented as causally autonomous when using decision theory, but argues that one-boxing is rational because the agent's intention should be construed as an uncaused cause of the predictor's prediction. If GIDT were applied given knowledge of the graph that Spohn favors, then, indeed, it would get the result that it is rational for the NP subject to one-box. But given that causes must precede their effects (as Spohn himself assumes), Spohn's DAG is inconsistent with the most natural reading of NP, according to which the agent must form the intention *now* to one-box or two-box (after the predictor has made her prediction).

[26] If the agent entertains DAGs according to which the NP subject's action exhibits some retrocausal influence on the predictor's prediction, then the combination of GIDT and the causal autonomy constraint can allow for dependencies between the intention and the prediction in live causal hypotheses. I set aside the possibility of retrocausality here.

outcome of a choice must seem open to the agent. It is at least somewhat plausible, for example, that while choosing whether to have peanut butter, I can entertain a causal hypothesis according to which my genetic condition exhibits causal control over what intention I form. Of course, if my choice must seem open to me, then I can't know that I will settle on having peanut butter (or on not having peanut butter) while making my choice. But there is a lot of conceptual space between an agent's not knowing that she will form some particular intention and an agent's regarding her intention as causally autonomous – conceptual space that can be used by philosophers to propose different constraints on intentions in order to capture the sense in which an agent's choice must be represented as open.

The weakest constraint that one might propose in order to account for this sense of openness is that an agent can decide whether to x only if she assigns positive subjective probability to her intending to x as she deliberates – or more formally, that $P(I = i_x) > 0$ for every x that is on the table.[27] If knowledge entails certainty, then this might capture the sense in which I cannot know whether I will x while deciding whether to x. But if this is the only constraint on intentions, then an agent's attitude towards her intention is not required to be epistemically free (or unconstrained by the agent's evidence about how she is likely to intend). For example, the NP subject could be 25% confident that she will intend to two-box as she chooses whether to two-box because the evidence that she has about her past brain states leads her to be this confident. Likewise, my evidence about my background and genetics could lead me to be 25% confident that I will settle on peanut butter as I choose whether to have peanut butter.

If this is the only constraint on intentions, then if the NP subject is very confident that she is not intervening (which is reasonable since most subjects must not intervene given that the predictor is remarkably successful), then one-boxing is the rational choice. Consider the results of applying GIDT. In order to determine the expected utility of one-boxing (or two-boxing), the agent must first determine how much she values intending to make herself one-box (or two-box) in each of the live causal hypotheses. This effectively amounts to determining how much she values the worlds that result from intending both options in each of the candidate chance distributions, where the effect of intending her action in each live chance distribution is the effect of conditioning on intending her action. In the distributions that are

---

[27] Or even weaker: that some probability distribution in the agent's credal state (which may consist of a non-singleton set of probability distributions) satisfies this constraint.

compatible with intervening, the chance of A's non-descendants must be equivalent to what they are unconditionally (since the CMC entails that I is not correlated with A's non-descendants). But in the live distributions that are compatible with not intervening – i.e., the distributions compatible with the DAG in Figure 4 – the chance distributions over P and B are highly affected by the NP subject's intention (given the predictor's success). Since the subject assigns the bulk of her subjective probability to causal hypotheses in which she does not intervene, and since one-boxing does so much better than two-boxing in each of these causal hypotheses, one-boxing beats two-boxing.

This suggests that one-boxers may champion one-boxing because they take the constraints on intentions to be relatively weak. That is, from the view of causal graphs, if an agent can make a choice between two options whenever she is uncertain which option she will intend, then one-boxing is favored over two-boxing because the agent should be very confident that she is not intervening when making her choice. Whether it is reasonable to adopt such a thin view of intentions is beyond the scope of this chapter, but if this view is reasonable, then, so, too, is one-boxing.

6 Epistemic Freedom without Causal Autonomy

I mentioned in the last two sections that many philosophers believe that an agent's doxastic attitudes towards how she will decide must be unconstrained by her evidence. The motivating idea behind this thought is that as an agent chooses what to do, she must think that what she does is up to her insofar as her doxastic attitudes towards what she does must not be constrained by evidence about factors outside of her control. Though this kind of "epistemic freedom" appears to be secured when the agent represents her intention as causally autonomous (since treating intentions as causally autonomous implies that they are uncorrelated with these factors in every causal hypothesis under consideration), it may be that there are contexts in which intentions are not represented as causally autonomous, but in which agents' attitudes towards their intentions are nevertheless unconstrained (in some sense) by their evidence about factors that lie beyond their control.

For example, even if the agent regarded her intention as *not* causally autonomous (as in Figure 4) and thus constrained by evidence about B or P were she to acquire it, it seems that the agent would qualify as epistemically free if she had not actually acquired any such evidence about B or P (because she would then lack the evidence that would constrain her relevant attitudes were she to have it). If this is right, then there should be a way of formulating constraints on agency that make no reference to causal autonomy and that imply that whether the agent has a genuine choice depends on whether the evidence that she has acquired permits her to be neutral with respect to how she will decide.

There is a long tradition of philosophers (e.g., Levi 1997; Price 2012; Spohn 1977) who attempt something like this by arguing that "deliberation crowds out prediction,"[28] or that agents cannot have subjective probabilities for how they will decide as they decide. Though the arguments for this position are many,[29] its advocates typically agree that deliberating agents should have no guess or "prediction" about how they will choose – that is, that the agent should be *neutral* with respect to how she will choose. What exactly this neutrality amounts to is not clear since these authors traditionally say nothing about what kind of doxastic attitude exists towards a proposition in the absence of subjective probability. But whatever it is, it seems that this line of reasoning preserves the intuition that deliberating agents are epistemically free. If the agent has no guess or prediction about how she will decide, then, *ipso facto*, she has no guess or prediction that is informed or constrained by her evidence.[30]

Since GIDT requires that probabilities be assigned to causal hypotheses (which themselves contain objective probability distributions over intention variables),[31] this view is a non-starter in the current context.[32] Moreover, as Hájek has recently (2016) argued, it is hard to understand why an agent's options should be perceived as blind spots (or impossible inputs) for subjective probability functions. But there may be some other way of requiring that agents be neutral towards how they will decide that does not bar agents from having subjective probabilities about how they will decide. That is, there may be some way to model agents as neutral towards their options that is consistent with agents having (some particular kind of) subjective probabilities over intention variables.

---

[28] This is Levi's (1997) way of describing the thesis.

[29] Following Spohn (1977), Levi (1997) argues that if the agent assigns probabilities to her options, the principles of rational choice will not apply non-vacuously.

[30] Price (2012) explicitly argues that agents are epistemically free in the sense that matters for agency when they do not have probabilities for how they will decide.

[31] If there are normative rules (e.g., Lewis's [1986] Principal Principle) that require the agent to have specific subjective probabilities over I given her subjective probabilities towards objective chance distributions over I, then the agent who applies GIDT should have subjective probabilities over I.

[32] But see Spohn (2012) for an effort to use causal graphs for decision-theoretic purposes while maintaining the spirit of this constraint.

In the context of GIDT, it seems that the agent is neutral towards how she will decide (in the sense that she has no guess or prediction about how she will decide) when she is neutral with respect to which of the possible unconditional chance distributions over I is correct. There may be multiple ways to represent neutrality of this sort, but it is at least reasonable to think of the agent as not having precise probability estimates about which unconditional chance distribution over I is true. For were the agent to be, say, 50% confident that the chance that she will intend to two-box is 20% and 50% confident that the chance that she will intend to two-box is 10%, then she should be 15% confident that she will opt to two-box.[33] More generally, if the agent is neutral towards how she will intend, then her confidence that $I = i_x$ is not representable as a precise probability estimate.[34]

One popular way to represent an agent as neutral towards some partition (e.g., the intention variable) is to use sets of probability functions (or "imprecise probabilities") to represent the agent's attitude. Levi (1974) argues, for example, that an agent might justifiably refrain from assigning each cell of a partition some unique probability when she lacks determinate evidence about where the truth lies within the partition, and that she might instead justifiably occupy a mental state that is best represented as a non-singleton convex set of probabilities over the partition.[35] In the extreme case, when an agent is *maximally* neutral towards some partition (perhaps because she has *no* evidence about where the truth lies), the agent can be best represented as a maximally inclusive set of probability functions – i.e., a set that jointly outputs every possible probability assignment for every cell in the partition (except perhaps extreme probability assignments of 1 and 0).[36] Though this line of reasoning requires more attention and development than it can be given here, this suggests that we might capture the sense in which agents should have no

---

[33] This follows only given that she rationally updates her subjective probability distribution in correspondence with her uncertainty about the chances.

[34] The agent likewise would not qualify as neutral (in the relevant sense) were she to have *equal* confidence that she one-boxes or two-boxes. For were the agent to be 50% confident that she'd one-box and 50% confident that she'd two-box, then she wouldn't be neutral about whether opting to one-box or two-box is more likely since she'd guess that it is no more likely that she does one or the other.

[35] See Levi (1974) for a classic discussion about why we would be moved to represent agents in this way, and Joyce (2010b) for a different approach to representing agents with imprecise probabilities.

[36] There may be reason to exclude extreme probability estimates in the context of decision-making since the agent perhaps *should* be able to rule out causal hypotheses according to which it is *certain* that she will form some particular intention – not only because entertaining these hypotheses does not jive with the phenomenology of choice, but also because the intention variable cannot be correlated with any other variables when one of its elements is assigned certainty (at least given the standard definition of conditional probability).

guess or prediction about how they will decide by requiring that the credal state of a deliberating agent be represented as a set of probability functions that jointly outputs every possible (non-extreme) probability estimate for every cell in the I partition.

What does this constraint mean for NP? Typically, when the details of NP are presented, the subject is not given any information that bears on the unconditional probability that she will one-box or two-box since she is not told how likely it is that the predictor predicts that she will one-box or two-box. So, regardless of whether her intention is causally autonomous (as in Figure 3) or endogenous (as in Figure 4), it seems reasonable to occupy a mental state that is best represented as maximally neutral towards how she will intend since she has no evidence that bears on how she will decide.[37] Thus it appears that, given this constraint, when GIDT is applied to the standard specification of NP, the expected utility calculation of one-boxing or two-boxing agrees with the last section's calculation, and the agent correspondingly discovers that she should one-box if she is reasonable.[38]

But if the details of the problem are changed such that the NP subject is told the unconditional chance that the predictor predicts that she will one-box – i.e., not only that the predictor is remarkably successful, but also that she predicts one-boxing at some particular rate – then the agent is forced to view her intention as an intervention when deliberating. For were the agent to entertain causal hypotheses according to which her intention is downstream from B (i.e., causal hypotheses that are partially comprised by the DAG in Figure 4), then the agent would have evidence about how she would decide,[39] and therefore could not reasonably adopt the maximally neutral set of

---

[37] This seems especially plausible since NP is usually presented to its subject out of thin air, and the NP subject correspondingly lacks any information about the predictor that goes beyond the details of the problem. Were the subject to know, for example, that the predictor was stingy, then she might have reason to believe to that she'd predict two-boxing more often than one-boxing (since the predictor loses less money when she predicts two-boxing), but since these details are not known by the NP subject, it is reasonable to think that the NP subject is totally in the dark about what the predictor will predict.

[38] Note that adopting an imprecise credal state towards I does not entail that the agent is imprecise towards any causal hypothesis if, as I advocate in Stern (2017), causal hypotheses can themselves be represented in terms of sets of objective chance distributions.

[39] When an agent has an unconditional probability estimate for A, and likewise has probability estimates for $P(A|B)$ and $P(B|A)$, her unconditional probability estimate for B is recoverable given the standard definition of conditional probability. Suppose, for example, that $P(A) = .1$, $P(A|B) = .4$, and $P(B|A) = .2$. Since $P(B|A) = P(A \& B) \div P(A)$, we can derive that $P(A \& B) = .02$. And since (i) $P(A|B) = P(A \& B) \div P(B)$, and (ii) we have values for $P(A|B)$ and $P(B)$, we can derive that $P(B) = .05$.

probabilities towards her options.[40] Thus the agent who is told how frequently the predictor predicts one-boxing (or two-boxing) is forced to think that she has a genuine decision to make only when she intervenes (because she otherwise cannot reasonably be maximally neutral towards how she will decide), and thus, when applying GIDT, discovers that she should two-box because her expected utility calculations reduce to those of IDT.[41]

If there are people who implicitly assume (i) that agents must be epistemically free and (ii) that agents need not represent themselves as causally autonomous, then these people's intuitions about NP should flip upon hearing the unconditional chance that the predictor has picked in a particular way. Indeed, it may be that many two-boxers implicitly make some guess about what the predictor will do, and that many one-boxers abstain from making any such guess. If this is right, and if it can be shown that people's reactions tend to flip when it is stipulated that the predictor will (or will not) predict one-boxing (or two-boxing) with some particular chance, then this would seem to be good news for the thesis that agents must be epistemically free, but need not be represented as causally autonomous when they choose. But whether people's reactions do, in fact, flip in this way is an empirical question that evades our grasp here.

Regardless, if whether an agent makes a genuine choice or forms a genuine intention is just a question of whether the agent is epistemically free, it seems that the solution to NP is slightly more complicated than if either of the two

[40] One might argue that the agent can be neutral about how she will decide even when she treats her intention as endogenous because the fact that she is making a decision gives her inadmissible evidence that trumps what she knows about the nature of the predictor's success, and correspondingly gives her reason to assign indeterminate probability to how she will decide. Though this argument cannot be immediately dismissed, there are two reasons to doubt its success. First, upon acquiring inadmissible evidence, we usually become *less* neutral about what will happen (because, e.g., the crystal ball tells us precisely what will happen), but, in order for the argument to go through, the agent must become *more* neutral upon acquiring the inadmissible evidence that she is making a choice. Second, even if the agent does acquire inadmissible evidence that rationally requires her to be neutral, it seems that she likewise learns that her intention is exogenous to the causal system at hand (because this is the only way to preserve the indeterminacy that is consistent with the axioms of the graphical approach to causal modeling). So it seems that this view reduces to some form of the view that intentions must be represented as causally autonomous – i.e., IDT.

[41] Though discussion of so-called "medical Newcomb Problems" evades our consideration here, the defender of this constraint can perhaps explain not only why the NP subject should one-box when confronted with the standard version of NP, but also why agents should not worry about backtracking non-causal dependencies between their action (e.g., smoking) and other symptoms when confronting medical Newcomb Problems. NP differs from realistic medical Newcomb Problems insofar as it is reasonable to lack an unconditional probability estimate for the predictor's prediction in NP, but not the correlated symptom (e.g., lung cancer) in realistic medical Newcomb Problems (because we have evidence that is relevant to whether we will get the correlated symptom in realistic cases).

previous constraints is right. If the subject has evidence about some variable correlated with her action, A – e.g., P or B – then she must view her intention as causally and evidentially independent from that variable, and should therefore two-box according to GIDT. But if she lacks evidence about these variables, then she is free to regard her intention as correlated with (and not d-separated from) that variable (as in Figure 4), and GIDT therefore licenses one-boxing (if her confidence that she is intervening is as low as it should be).

7 Conclusion

We have now seen that three views about the nature of choice yield different consequences for NP when combined with GIDT. If agents must represent themselves as causally autonomous when they make choices, then NP subjects should two-box no matter how the details of NP are further specified. If agents need not represent themselves as causally autonomous nor be epistemically free, and if agents must instead only assign positive probability to their intending every option on the table, then NP subjects should one-box provided that the predictor is known to be remarkably successful. But if agents must be epistemically free in order to make choices and need not represent themselves as causally autonomous, then NP subjects should two-box when they are provided with information that is relevant to the unconditional probability that the predictor predicts one-boxing (or two-boxing), but should otherwise one-box (because their epistemic freedom is consistent with regarding their intention as correlated with the predictor's prediction when they have no evidence about what the predictor will do).

So, what should NP subjects do? In order to answer this question, we must settle the very difficult question of how agents must represent themselves when making decisions, and this is not a matter that we can settle here. But by tracing the consequences of assuming different answers to this question from within the graphical approach to causal modeling, I hope to have presented a new framework for investigating the solution to NP.

References

Abramson, P. R., J. H. Aldrich, P. Paolino and D. W. Rohde. 1992. 'Sophisticated' voting in the 1988 Presidential Primaries. *American Political Science Review* 86: 55–69.

Ahmed, A. 2007. Agency and causation. In Corry, R. and H. Price (ed.), *Causation, Physics and the Constitution of Reality: Russell's Republic Revisited*. Oxford: Oxford University Press: 120–55.

Ahmed, A. 2014a. *Evidence, Decision and Causality*. Cambridge: Cambridge University Press.

Ahmed, A. 2014b. Dicing with death. *Analysis* 74: 587–92.

Ahmed, A. 2015. Infallibility in the Newcomb problem. *Erkenntnis* 80: 261–73.

Ahmed, A. 2017. Exploiting Causal Decision Theory. Manuscript.

Ahmed, A. Forthcoming. Self-control and hyperbolic discounting. In Bermúdez, J. L. (ed.) *Self-Control & Rationality: Interdisciplinary Essays*. Cambridge: Cambridge University Press.

Ahmed, A. and A. Caulton. 2014. Causal Decision Theory and EPR correlations. *Synthese* 191: 4315–52.

Ainslie, G. 1991. Derivation of 'rational' economic behaviour from hyperbolic discount curves. *American Economic Review* 81: 334–40.

Al-Najjar, N. and J. Weinstein. 2009. The ambiguity aversion literature: a critical assessment. *Economics and Philosophy* 25: 249–84.

Allais, M. 1953. Le comportement de l'homme rationnel devant le risque: critique des postulats et axiomes de l'école Américaine. *Econometrica* 21: 503–46.

Alston, W. P. 1988. An internalist externalism. *Synthese* 74: 265–83.

Alston, W. P. 2005. *Beyond 'Justification': Dimensions of Epistemic Evaluation*. Ithaca: Cornell University Press.

Andreou, C. 2008. The Newxin Puzzle. *Philosophical Studies* 139: 415–22.

Arntzenius, F. 2003. Some problems for conditionalization and reflection. *Journal of Philosophy* 100: 356–70.

Arntzenius, F. 2008. No regrets, or: Edith Piaf revamps decision theory. *Erkenntnis* 68: 277–97.

Arntzenius, F., A. Elga and J. Hawthorne. 2004. Bayesianism, infinite decisions, and binding. *Mind* 113: 251–83.

Aumann, R. 1987. Correlated equilibrium as an expression of Bayesian rationality. *Econometrica* 55: 1–18.

Aumann, R. and A. Brandenburger. 1995. Epistemic conditions for Nash equilibrium. *Econometrica* 63: 1161–80.

Balleine B. W., A. Espinet and F. Gonzalez. 2005. Perceptual learning enhances retrospective revaluation of conditioned flavour preferences in rats. *Journal of Experimental Psychology: Animal Learning and Cognition* 31: 341–50.

Baratgin, J. and G. Politzer. 2010. Updating: a psychologically basic situation of probability revision. *Thinking & Reasoning* 16: 253–87.

Battigalli, P. and G. Bonanno 1999. Recent results on belief, knowledge and the epistemic foundations of game theory. *Research in Economics* 53: 149–225.

Beebee, H. and D. Papineau. 1997. Probability as a guide to life. *Journal of Philosophy* 94: 217–43.

Berkeley, G. 1980 [1710]. *Principles of Human Knowledge*. In his *Philosophical Works*, Ayers, M. (ed.). London: Everyman.

Bermúdez, J. L. 2009. *Decision Theory and Rationality*. Oxford: Oxford University Press.

Bermúdez, J. L. 2013. Prisoner's dilemma and Newcomb's problem: why Lewis's argument fails. *Analysis* 73: 423–29.

Bermúdez, J. L. 2015a. Prisoner's dilemma cannot be a Newcomb problem. In Peterson, M. (ed.), *The Prisoner's Dilemma*. Cambridge: Cambridge University Press: 115–32.

Bermúdez, J. L. 2015b. Strategic vs. parametric choice in Newcomb's Problem and the Prisoner's Dilemma: reply to Walker. *Philosophia* 43: 787–94.

Binmore, K. 1988. Modeling rational players: part II. *Economics and Philosophy* 4: 9–55.

Binmore, K. 1993. De-Bayesing game theory. In Binmore, K., A. Kirman and P. Tani (ed.), *Frontiers of Game Theory*. Cambridge, MA: MIT Press: 321–40.

Bradley, R. 1998. A representation theorem for a decision theory with conditionals. *Synthese* 116: 187–229.

Brandenburger, A. 1992. Knowledge and equilibrium in games. *Journal of Economic Perspectives* 6: 83–101.

Bratman, M. E. 1999. *Intentions, Plans, and Practical Reason*. Stanford: CSLI Publications.

Briggs, R. A. 2010. Decision-theoretic paradoxes as voting paradoxes. *Philosophical Review* 119: 1–30.

Broome, J. 1989. An economic Newcomb problem. *Analysis* 49: 220–22.

Broome, J. 1990a. Bolker-Jeffrey expected utility theory and axiomatic utilitarianism. *Review of Economic Studies* 57: 477–502.

Broome, J. 1990b. Should a rational agent maximize expected itility? In Schweers Cook, Karen and Margaret Levi (ed.), *The Limits of Rationality*. Chicago: University of Chicago Press: 132–45.

Broome, J. 1991a. *Weighing Goods*. Oxford: Basil Blackwell.

Broome, J. 1991b. Rationality and the Sure Thing Principle. In Meeks, J. G. T. (ed.), *Thoughtful Economic Man*. Cambridge: Cambridge University Press: 74–102.

Broome, J. 2001. Are intentions reasons? And how should we cope with incommensurable values? In Morris, C. and A. Ripstein (ed.), *Practical Rationality and Preference: Essays for David Gauthier*. Cambridge: Cambridge University Press: 98–120.

Byrne, A. and A. Hájek. 1997. David Hume, David Lewis, and decision theory. *Mind* 106: 411–28.

Cain, B. E. 1978. Strategic voting in Great Britain. *American Journal of Political Science* 22: 639–55.

Cartwright, N. 1979. Causal laws and effective strategies. *Noûs* 13: 419–37.

Cavalcanti, E. 2010. Causation, decision theory and Bell's theorem: a quantum analogue of the Newcomb problem. *British Journal for the Philosophy of Science* 61: 569–97.

Charig, C. R., D. R. Webb, S. R. Payne and O. E. Wickham. 1986. Comparison of treatment of renal calculi by operative surgery, percutaneous nephrolithotomy, and extracorporeal shock wave lithotripsy. *British Medical Journal* 292: 879–82.

Davidson, D. 2001. *Inquiries into Truth and Interpretation*. Oxford: Clarendon Press.

Dawes, R. M. 1990. The potential nonfalsity of the false consensus effect. In Hogarth, R. M. (ed.), *Insight in Decision Making*. Chicago: University of Chicago Press: 179–99.

Dickinson, A. 2012. Associative learning and animal cognition. *Philosophical Transactions of the Royal Society B: Biological Sciences* 367: 2733–42.

Dennett, D. 1984. *Elbow Room: The Varieties of Free Will Worth Wanting*. Cambridge, MA: MIT Press.

Dennett, D. 1987. *The Intentional Stance*. Cambridge, MA: MIT Press.

Dennett, D. 2003. *Freedom Evolves*. New York: Viking.

Dennett, D. 2015. *Elbow Room: The Varieties of Free Will Worth Wanting*. 2nd ed. Cambridge, MA: MIT Press.

Dummett, M. A. E. 1954. Can an effect precede its cause? *Proceedings of the Aristotelian Society, Supplementary Volume* 28: 27–44.

Dummett, M. A. E. 1964. Bringing about the past. *Philosophical Review* 73: 338–59. Reprinted in his *Truth and Other Enigmas*. London: Duckworth: 333–50.

Dummett, M. A. E. 1987. Reply to D. H. Mellor. In Taylor, B. (ed.), *Michael Dummett: Contributions to Philosophy*. Dordrecht: Martinus Nijhoff: 287–98.

Dummett, M. A. E. 1986. Causal loops. In Flood, R. and M. Lockwood (ed.), *The Nature of Time*. Oxford: Blackwell: 135–69. Reprinted in Dummett, M. A. E., *The Seas of Language*. Oxford: Oxford University Press 1993: 349–75.

Downs, A. 1957. *An Economic Theory of Democracy*. New York: Harper & Row.

Eells, E. 1981. Causality, utility, and decision. *Synthese* 48: 295–329.

Eells, E. 1982. *Rational Decision and Causality*. Cambridge: Cambridge University Press.

Eells, E. 1984. Newcomb's many solutions. *Theory and Decision* 16: 59–105.

Eells, E. 1985. Causality, decision, and Newcomb's paradox. In Campbell, R. and L. Sowden (ed.), *Paradoxes of Rationality and Cooperation: Prisoner's Dilemma and Newcomb's Problem*. Vancouver: University of British Columbia Press: 183–213.

Eells, E. and W. Harper. 1991. Ratifiability, game theory, and the principle of independence of irrelevant alternatives. *Australasian Journal of Philosophy* 69: 1–19.

Egan, A. 2007. Some counterexamples to Causal Decision Theory. *Philosophical Review* 116: 93–114.

El Skaf, R. and C. Imbert. 2013. Unfolding in the empirical sciences: experiments, thought experiments and computer simulations. *Synthese* 190: 3451–74.

Ellsberg, D. 1961. Risk, ambiguity and the Savage axioms. *Quarterly Journal of Economics* 75: 643–69.

Elster, J. 1989. *Solomonic Judgements*. Cambridge: Cambridge University Press.

Field, H. H. 1977. Logic, meaning, and conceptual role. *Journal of Philosophy* 74: 379–409.

Fiorina, M. P. 1990. Information and rationality in elections. In Ferejohn, J. A. and J. H. Kuklinski (ed.), *Information and Democratic Processes*. Urbana: University of Illinois Press: 329–42.

Fischer, A. J. 1999. The probability of being decisive. *Public Choice* 101: 267–83.

Fishburn, P. C. 1964. *Decision and Value Theory*. New York: John Wiley & Sons.

Fiske, S. T. and S. E. Taylor. 1984. *Social Cognition*. Reading, MA: Addison-Wesley.

Fleurbaey, M. and A. Voorhoeve. 2013. Decide as you would with full information! An argument against *ex ante* Pareto. In Eyal, N., S. A. Hurst, O. F. Norheim and D. Wikler (ed.), *Inequalities in Health: Concepts, Measures, and Ethics*. Oxford: Oxford University Press: 113–28.

Friedman, M. 1953. *Essays in Positive Economics*. Chicago: University of Chicago Press.

Frydman, R., G. P. O'Driscoll and A. Schotter. 1982. Rational expectations of government policy: an application of Newcomb's problem. *Southern Economic Journal* 1982: 311–19.

Fudenberg, D. and J. Tirole. 1991. *Game Theory*. Cambridge, MA: MIT Press.

Gauthier, D. 1986. *Morals by Agreement*. Oxford: Oxford University Press.

Gauthier, D. 1989. In the neighbourhood of the Newcomb-Predictor. *Proceedings of the Aristotelian Society* 89: 179–94.

Gauthier, D. 1994. Assure and threaten. *Ethics* 194: 690–721.

Gibbard, A. 1992. Weakly self-ratifying strategies: comments on McClennen. *Philosophical Studies* 65: 217–25.

Gibbard, A. and W. Harper. 1978. Counterfactuals and two kinds of expected utility. In Hooker, A., J. J. Leach and E. F. McClennen (ed.), *Ifs: Conditionals, Belief, Decision, Chance and Time*. Dordrecht: D. Reidel: 153–90.

Good, I. J. 1967. On the principle of total evidence. *British Journal for the Philosophy of Science* 17: 319–21.

Grafstein, R. 1991. An Evidential Decision Theory of turnout. *American Journal of Political Science* 35: 989–1010.

Grafstein, R. 1999. *Choice-Free Rationality*. Ann Arbor: University of Michigan Press.

Grafstein, R. 2002. What rational political actors can expect. *Journal of Theoretical Politics* 14: 139–65.

Grafstein, R. 2003. Strategic voting in presidential primaries: problems of explanation and interpretation. *Political Research Quarterly* 56: 513–19.

Greene, P. 2013. *Rationality and Success*. Dissertation. New Brunswick: Rutgers, The State University of New Jersey.

Hájek, A. 2016. Deliberation welcomes prediction. *Episteme* 13: 507–28.

Hájek, A. A Puzzle about Degree of Belief. Manuscript.

Hall, N. 1994. Correcting the guide to objective chance. *Mind* 103: 505–18.

Hall, N. 2004. Two mistakes about credence and chance. *Australasian Journal of Philosophy* 82: 93–111.

Hare, C. and B. Hedden. 2016. Self-reinforcing and self-frustrating decisions. *Noûs* 50: 604–28.

Harper, W. L. 1975. Rational belief change, Popper functions and counterfactuals. *Synthese* 30: 221–62.

Harper, W. L. 1986. Mixed strategies and ratifiability in Causal Decision Theory. *Erkenntnis* 24: 25–36.

Hedden, B. 2015. Time-slice rationality. *Mind* 124: 449–91.

Hesslow, G. 1976. Discussion: two notes on the probabilistic approach to causality. *Philosophy of Science* 43: 290–2.

Hitchcock, C. 1996. Causal Decision Theory and decision-theoretic causation. *Noûs* 30: 508–26.

Hitchcock, C. 2015. Conditioning, intervening and decision. *Synthese* 193: 1157–76.

Horgan, T. 1981. Counterfactuals and Newcomb's problem. *Journal of Philosophy* 78: 331–56.

Horgan, T. 1985. Newcomb's problem: a stalemate. In Campbell, R. and L. Sowden (ed.), *Paradoxes of Rationality and Co-operation: Prisoner's Dilemma and Newcomb's Problem.* Vancouver: University of British Columbia Press: 223–33.

Horwich, P. 1987. *Asymmetries in Time: Problems in the Philosophy of Science.* Cambridge, MA: MIT Press.

Hurley, S. L. 1991. Newcomb's problem, prisoners' dilemma, and collective action. *Synthese* 86: 173–96.

Hurley, S. L. 1994. A new take from Nozick on Newcomb's Problem and Prisoners' Dilemma. *Analysis* 54: 65–72.

Ismael, J. 2007. Freedom, compulsion, and causation. *Psyche* 13: 1–10.

Jackson, F. and R. Pargetter. 1983. Where the tickle defence goes wrong. *Australasian Journal of Philosophy* 61: 295–99.

Jacobi, N. 1993. Newcomb's paradox: a realist resolution. *Theory and Decision* 35: 1–17.

James, W. 1912. The will to believe. In his *Will to Believe and Other Essays in Popular Philosophy.* New York: Henry Holt and Co.: 1–32. Reprinted in McDermott, J. J. (ed.), *The Writings of William James.* New York: Random House. Reprint, Chicago: University of Chicago Press 1978: 717–35.

Jeffrey, R. C. 1965. *The Logic of Decision.* Chicago: University of Chicago Press.

Jeffrey, R. C. 1983. *The Logic of Decision.* 2nd ed. Chicago: University of Chicago Press.

Joyce, J. 1998. A nonpragmatic vindication of probabilism. *Philosophy of Science* 65: 575–603.

Joyce, J. 1999. *Foundations of Causal Decision Theory.* Cambridge: Cambridge University Press.

Joyce, J. 2002. Levi on Causal Decision Theory and the possibility of predicting one's own actions. *Philosophical Studies* 110: 69–102.

Joyce, J. 2007. Are Newcomb problems really decisions? *Synthese* 156: 537–62.

Joyce, J. 2010a Causal reasoning and backtracking. *Philosophical Studies* 147: 139–54.

Joyce, J. 2010b. A defence of imprecise credences in inference and decision making. *Philosophical Perspectives* 19: 153–78.

Joyce, J. 2012. Regret and instability in Causal Decision Theory. *Synthese* 187: 123–45.

Joyce, J. 2016. Review essay: Arif Ahmed, *Evidence, Decision and Causality. Journal of Philosophy* 113: 224–32.

Kahneman, D. and A. Tversky. 1984. Choices, values, and frames. *American Psychologist* 39: 341–50.

Kanazawa, S. 2000. A new solution to the collective action problem: the paradox of voter turnout. *American Sociological Review* 65: 433–42.

Kane, R. 1996. *The Significance of Free Will*. New York: Oxford University Press.

Katsuno, H. and A. O. Mendelzon. 1991. On the difference between updating a knowledge base and revising it. In *Proceedings of the Second International Conference on Principles of Knowledge Representation and Reasoning*. Vol. 2. San Francisco: Morgan Kaufman Publishers: 387–94.

Kavka, G. S. 1983. The Toxin Puzzle. *Analysis* 43: 33–36.

Kneale, W. C. 1949. *Probability and Induction*. Oxford: Clarendon Press.

Lange, M. 1999. Calibration and the epistemological role of Bayesian conditionalization. *Journal of Philosophy* 96: 294–324.

Ledyard, J. O. 1984. The pure theory of large two-candidate elections. *Public Choice* 44: 7–41.

Ledyard, J. O. 1995. Public goods: a survey of experimental research. In Kagel, J. H., and A. E. Roth (ed.), *The Handbook of Experimental Economics*. Princeton: Princeton University Press: 111–94.

LeRoy, S. F. 1995. On policy regimes. In Hoover, K. D. (ed.), *Macroeconometrics*. Boston: Kluwer: 235–61.

Levi, I. 1974. On indeterminate probabilities. *Journal of Philosophy* 71: 391–418.

Levi, I. 1975. Newcomb's many problems. *Theory and Decision* 6: 161–75.

Levi, I. 1989. Rationality, prediction and autonomous choice. *Canadian Journal of Philosophy* Supplemental Vol. 19: 339–63.

Levi, I. 1992. Feasibility. In Bicchieri, C. and M. L. Dalla Chiara (ed.), *Knowledge, Belief, and Strategic Interaction*. Cambridge: Cambridge University Press: 1–20.

Levi, I. 1997a. Advising rational agents. *Economics & Politics* 9: 221–24.

Levi, I. 1997b. *The Covenant of Reason: Rationality and the Commitments of Thought*. Cambridge: Cambridge University Press.

Levi, I. 2000. Review: James M. Joyce, *The Foundations of Causal Decision Theory*. *Journal of Philosophy* 97: 387–402.

Levi, I. 2012. *Hard Choices*. Cambridge: Cambridge University Press.

Lewis, D. K. 1973. *Counterfactuals*. Oxford: Blackwell.

Lewis, D. K. 1976a. The paradoxes of time travel. *American Philosophical Quarterly* 13:145–152. Reprinted in his *Philosophical Papers Vol. II*. Oxford: Oxford University Press, 1986: 67–80.

Lewis, D. K. 1976b. Probabilities of conditionals and conditional probabilities. *Philosophical Review* 85: 297–315.

Lewis, D. K. 1979. Prisoners' dilemma is a Newcomb problem. *Philosophy and Public Affairs* 8: 235–40.

Lewis, D. K. 1980. A subjectivist's guide to objective chance. In Jeffrey, R. C. (ed.), *Studies in Inductive Logic and Probability Vol. II*. Berkeley: University of California Press: 263–94. Reprinted in his *Philosophical Papers Vol. II*. Oxford: Oxford University Press 1986: 83–132.

Lewis, D. K. 1981a. Causal Decision Theory. *Australasian Journal of Philosophy* 59: 5–30. Reprinted in his *Philosophical Papers Vol. II*. Oxford: Oxford University Press, 1986: 305–39.

Lewis, D. K. 1981b. Why ain'cha rich? *Noûs* 15: 377–80. Reprinted in his *Papers in Ethics and Social Philosophy*. Cambridge: Cambridge University Press, 2000: 37–41.

Lewis, D. K. 1994a. Reduction of mind. In Guttenplan, S. (ed.), *A Companion to the Philosophy of Mind*. Oxford: Blackwell: 412–31.

Lewis, D. K. 1994b. Humean supervenience debugged. *Mind* 103: 473–90.

Lewis, D. K. 1999. Why Conditionalize? In his *Papers in Metaphysics and Epistemology*. Cambridge: Cambridge University Press: 403–7.

Lindley, D. and M. Novick. 1981. The role of exchangeability in inference. *Annals of Statistics* 9: 45–58.

Luce, R. D. and H. Raiffa. 1957. *Games and Decisions: Introduction and Critical Survey*. New York: John Wiley & Sons.

Mackie, J. L. 1977. Newcomb's Paradox and the direction of causation. *Canadian Journal of Philosophy* 7: 213–25.

Mackie, J. L. 1980. *The Cement of the Universe*. Oxford: Clarendon Press.

McClennen, E. F. 1990. *Rationality and Dynamic Choice*. Cambridge: Cambridge University Press.

McClennen, E. F. 1992. The theory of rationality for ideal games. *Philosophical Studies* 65: 193–215.

Meacham, C. J. G. 2010. Binding and its consequences. *Philosophical Studies* 149: 49–71.

Meek, C. and C. Glymour. 1994. Conditioning and intervening. *British Journal for the Philosophy of Science* 45: 1001–21.

Mehta, J., C. Starmer and R. Sugden. 1994. The nature of salience: an experimental investigation of pure coordination games. *American Economic Review* 84: 658–73.

Mellor, D. H. 1971. *The Matter of Chance*. Cambridge: Cambridge University Press.

Mellor, D. H. 1991. Objective decision making. In his *Matters of Metaphysics*. Cambridge: Cambridge University Press: 269–87.

Menzies, P. and H. Price. 1993. Causation as a secondary quality. *British Journal for the Philosophy of Science* 44: 187–203.

Meyer, D., J. Feldmaier and H. Shen. 2016. Reinforcement learning in conflicting environments for autonomous vehicles. *Conference proceedings: Robotics in the 21st Century: Challenges and Promises*. arXiv: 1610.07089v1.

Monterosso, J. and G. Ainslie. 1999. Beyond discounting: Possible experimental models of impulse control. *Psychopharmacology* 146: 339–47.

Moss, S. 2012. Updating as communication. *Philosophy and Phenomenological Research* 85: 225–48.

Moss, S. 2013. Epistemology formalized. *Philosophical Review* 122: 1–43.

Moss, S. Forthcoming. *Probabilistic Knowledge*. Oxford: Oxford University Press.

Nash, J. 1951. Non-cooperative games. *Annals of Mathematics* 54: 286–95.

Nietzsche, F. 2001 [1886]. *Beyond Good and Evil*. Tr. J. Norman. Ed. R.-P. Horstmann. Cambridge: Cambridge University Press.

Nozick, R. 1969. Newcomb's problem and two principles of choice. In N. Rescher (ed.), *Essays in Honor of Carl G. Hempel*. Dordrecht: D. Reidel: 114–46. Reprinted in P. Moser (ed.), *Rationality in Action: Contemporary Approaches*. Cambridge: Cambridge University Press, 1990: 207–34.

Nozick, R. 1981. *Philosophical Explanations*. Cambridge, MA: Harvard University Press.

Nozick, R. 1993. *The Nature of Rationality*. Princeton: Princeton University Press.

O'Flaherty, B. and J. Bhagwati. 1997. Will free trade with political science put normative economists out of work? *Economics & Politics* 9: 207–19.

Okasha, S. 2015. On the interpretation of decision theory. *Economics and Philosophy* 32: 409–33.

Orbell, J. M., A. J. C. van de Kragt, and R. M. Dawes. 1991. Covenants without the sword: the role of promises is social in social dilemma circumstances. In Koford, K. J. and J. B. Miller (ed.), *Social Norms and Economic Institutions*. Ann Arbor: University of Michigan Press: 117–34.

Palfrey, T. R. and H. Rosenthal. 1985. Voter participation and strategic uncertainty. *American Political Science Review* 79: 62–78.

Parfit, D. 1984. *Reasons and Persons*. Oxford: Oxford University Press.

Pearl, J. 2000. *Causality: Models, Reasoning, and Inference*. Cambridge: Cambridge University Press.

Pearl, J. 2009. *Causality: Models, Reasoning, and Inference*. 2nd ed. Cambridge: Cambridge University Press.

Pearl, J. 2014. Understanding Simpson's Paradox. *American Statistician* 68: 8–13.

Pettit, P. 1988. The prisoner's dilemma is an unexploitable Newcomb problem. *Synthese* 76: 123–34.

Pettit, P. 1991. Decision theory and folk psychology. In Bacharach, M. and S. L. Hurley (ed.), *Foundations of Decision Theory: Issues and Advances*. Oxford: Basil Blackwell: 147–75.

Pollock, J. L. 2006. *Thinking about Acting: Logical Foundations for Rational Decision Making*. Oxford: Oxford University Press.

Price, H. 1986. Against causal decision theory. *Synthese* 67: 195–212.

Price, H. 1992. The direction of causation: Ramsey's ultimate contingency. *PSA: Proceedings of the Biennial Meeting of the Philosophy of Science Association*. East Lansing, MI: Philosophy of Science Association: 253–67.

Price, H. 2001. Causation in the special sciences: the case for pragmatism. In Costantini, D., M. C. Galavotti and P. Suppes (ed.), *Stochastic Causality*. Stanford: CSLI Publications: 103–20.

Price, H. 2007. Causal perspectivalism. In Corry, R. and H. Price (ed.), *Causation, Physics and the Constitution of Reality: Russell's Republic Revisited*. Oxford: Oxford University Press: 250–92.

Price, H. 2012. Causation, chance, and the rational significance of supernatural evidence. *Philosophical Review* 121: 483–538.

Quattrone, G. A. and A. Tversky. 1986. Self-deception and the voter's illusion. In Elster, J. (ed.), *The Multiple Self*. Cambridge: Cambridge University Press: 35–58.

Quattrone, G. A. and A. Tversky. 1988. Contrasting rational and psychological analyses of political choice. *American Political Science Review* 82: 719–36.

Quinn, W. 1990. The puzzle of the selftorturer. *Philosophical Studies* 59: 79–90. Reprinted in his *Morality and Action*. Cambridge: Cambridge University Press 1993: 198–209.

Rabinowicz, W. 2002. Does practical deliberation crowd out self-prediction? *Erkenntnis* 57: 91–122.

Ramsey, F. P. 1990 [1926]. Truth and probability. In his *Philosophical Papers*, ed. D. H. Mellor. Cambridge: Cambridge University Press, 1990: 52–94.

Ramsey, F. P. 1978 [1929]. General propositions and causality. In Mellor, D. H. (ed.), *Foundations: Essays in Philosophy, Logic, Mathematics and Economics*. London: Routledge: 133–51.

Reichenbach, H. 1956. *The Direction of Time*. Berkeley: University of California Press.

Resnik, M. 1987. *Choices: An Introduction to Decision Theory*. Minneapolis: University of Minnesota Press.

Sargent, T. J. 1981. Rational expectations and the theory of economic policy. In Lucas, R. E. Jr and T. J. Sargent (ed.), *Rational Expectations and Econometric Practice*. Minneapolis: University of Minnesota Press: 199–214.

Sargent, T. J. 1993. *Bounded Rationality in Macroeconomics*. Oxford: Oxford University Press.

Sargent, T. J. and N. Wallace. 1976. Rational expectations and the theory of economic policy. *Journal of Monetary Economics* 2: 169–83.

Savage, L. J. 1972. *The Foundations of Statistics*. 2nd ed. New York: Dover Press.

Schelling, T. C. 1960. *The Strategy of Conflict*. Oxford: Oxford University Press.

Selten, R. 1975. Re-examination of the perfectness concept for equilibrium points in extensive games. *International Journal of Game Theory* 4: 22–55.

Seidenfeld, T. 1985. Comments on Causal Decision Theory. *PSA 1984* 2: 201–12.

Shoham, Y. and K. Leyton-Brown. 2009. *Multiagent Systems: Algorithmic, Game-Theoretic, and Logical Foundations*. Cambridge: Cambridge University Press.

Sims, C. A. 1986. Are forecasting models usable for policy analysis? *Federal Reserve Bank of Minneapolis Quarterly Review* 10: 2–16.

Sims, C. A. 2003. Implications of rational inattention. *Journal of Monetary Economics* 50: 665–90.

Skyrms, B. 1980. *Causal Necessity: A Pragmatic Investigation of the Necessity of Laws*. New Haven: Yale University Press.

Skyrms, B. 1982. Causal Decision Theory. *Journal of Philosophy* 79: 695–711.

Skyrms, B. 1990a. The value of knowledge. *Minnesota Studies in the Philosophy of Science* 14: 245–66.

Skyrms, B. 1990b. *The Dynamics of Rational Deliberation*. Cambridge, MA: Harvard University Press.

Slezak, P. 2013. Realizing Newcomb's problem. Available online at philsci-archive.pitt.edu/9634/1/Realizing_Newcomb's_Problem.pdf (accessed 19 December 2017).

Smith, H. 1991. Deriving Morality from Rationality. In Vallentyne, P. (ed.), *Contractarianism and Rational Choice: Essays on David Gauthier's Morals by Agreement*. Cambridge: Cambridge University Press: 229–53.

Soares, N. and B. Fallenstein. 2015. Toward idealized decision theory. *arXiv: 1507.01986 [cs.AI]*.

Soares, N., B. Fallenstein and B. Levinstein. Cheating Death in Damascus. Available online at https://intelligence.org/files/DeathInDamascus.pdf. Manuscript.

Soares, N., B. Fallenstein and E. Yudkowsky. Functional Decision Theory: A New Theory of Instrumental Rationality. Manuscript.

Sobel, J. H. 1985a. Not every prisoner's dilemma is a Newcomb problem. In Campbell, R. and L. Sowden (ed.), *Paradoxes of Rationality and Cooperation*. Vancouver: University of British Columbia Press: 263–74.

Sobel, J. H. 1985b Circumstances and dominance in a Causal Decision Theory. *Synthese* 63: 167–202.

Sobel, J. H. 1986. Notes on decision theory. *Australasian Journal of Philosophy* 64: 407–37. Reprinted in his *Taking Chances*. Cambridge: Cambridge University Press 1994: 141–73.

Spencer, J. and I. Wells. 2017. Why take both boxes? *Philosophy and Phenomenological Research*. doi: 10.1111/phpr.12466.

Spirtes, P., C. Glymour and R. Scheines. 2000. *Causation, Prediction, and Search*. 2nd ed. Cambridge, MA: MIT Press.

Spohn, W. 1977. Where Luce and Krantz do really generalize Savage's decision model. *Erkenntnis* 11: 113–34.

Spohn, W. 2012. Reversing 30 years of discussion: why Causal Decision Theorists should one-box. *Synthese* 187: 95–122.

Stalnaker, R. C. 1972. Letter to David Lewis. Reprinted in Harper, W. L., R. Stalnaker and G. Pearce (ed.), *Ifs: Conditionals, Beliefs, Chance and Time*. Dordrecht: Springer, 1981: 151–52.

Stalnaker, R. C. 1996. Knowledge, belief and counterfactual reasoning in games. *Economics and Philosophy* 12: 133.

Stalnaker, R. C. 1997. On the evaluation of solution concepts. In Bacharach, M., L.-A. Gérard-Varet, P. Mongin and H. S. Shin (ed.), *Epistemic Logic and the Theory of Games and Decisions*. Boston: Kluwer Academic Publishers: 345–64.

Stalnaker, R. C. 1999. Extensive and strategic forms: games and models for games. *Research in Economics* 53: 293–319.

Stern, R. 2017. Interventionist decision theory. *Synthese* 194: 4133–53.

Strawson, G. 1986. *Freedom and Belief*. Oxford: Clarendon Press.

Strawson, G. 2003. Evolution explains it all for you. *New York Times* March 2. Available online at www.nytimes.com/2003/03/02/books/evolution-explains-it-all-for-you.html.

Suddendorf, T. and M. C. Corballis. 2007. The evolution of foresight: what is mental time travel, and is it unique to humans? *Behavioural and Brain Sciences* 30: 299–313.

Sugden, R. 1995. A theory of focal points. *Economic Journal* 105: 533–50.

Sugden, R. and I. E. Zamarrón. 2006. Finding the key: the riddle of focal points. *Journal of Economic Psychology* 27: 609–21.

Tang, W. H. 2016. Reliability theories of justified credence. *Mind* 125: 63–94.

Tawney, R. H. 1998 [1926]. *Religion and the Rise of Capitalism*. London: Routledge.

Teller, P. 1973. Conditionalization and observation. *Synthese* 26: 218–58.

Van Fraassen, B. 1983. Calibration: a frequency justification for personal probability. In Cohen, R. and L. Laudan (ed.), *Physics, Philosophy, and Psychoanalysis*. Dordrecht: D. Reidel: 295–319.

Van Fraassen, B. 1984. Belief and the will. *Journal of Philosophy* 81: 235–56.

Velleman, J. D. 1989. Epistemic freedom. *Pacific Philosophical Quarterly* 70: 73–97.

Von Neumann, J. and O. Morgenstern. 1944. *Theory of Games and Economic Behaviour*. Princeton: Princeton University Press.

Walker, M. T. 2014. The real reason why the Prisoner's dilemma is not a Newcomb problem. *Philosophia* 42: 841–59.

Weber, M. 1992 [1920]. *The Protestant Ethic and the Spirit of Capitalism*. Tr. T. Parsons. London: Routledge.

Wedgwood, R. 2013. Gandalf's solution to the Newcomb problem. *Synthese* 190: 2643–75.

Weirich, P. 1985. Decision Instability. *Australasian Journal of Philosophy* 63: 465–72.

Weirich, P. 2004. *Realistic Decision Theory: Rules for Nonideal Agents in Nonideal Circumstances*. Oxford: Oxford University Press.

Wells, I. Forthcoming. Equal opportunity and Newcomb's problem. *Mind*.

Woodward, J. 2003. *Making Things Happen*. Oxford: Oxford University Press.

Yudkowsky, E. 2010. Timeless decision theory. Technical Report, Machine Intelligence Research Institute (MIRI): www.intelligence.org.

Index